Hazardous Waste and Pollution

Tanya Wyatt
Editor

Hazardous Waste and Pollution

Detecting and Preventing Green Crimes

Springer

Editor
Tanya Wyatt
Northumbria University
Newcastle-upon-Tyne
Northumberland
UK

ISBN 978-3-319-18080-9 ISBN 978-3-319-18081-6 (eBook)
DOI 10.1007/978-3-319-18081-6

Library of Congress Control Number: 2015947130

Springer Cham Heidelberg New York Dordrecht London

Printed on acid-free paper

Springer International Publishing AG Switzerland is part of Springer Science+Business Media (www.springer.com)

Acknowledgements

I, and the contributing authors, would like to thank the UK Economic and Social Research Council for funding the Green Criminology Research Seminar Series. The chapters here were all presentations at the 'Brown Crime' seminar that was part of the series. Also, we would like to thank Northumbria University for the additional funding to make the seminar an all-day conference with speakers from around the world.

Contents

About the Editor

Tanya Wyatt is a Principal Lecturer in Criminology at Northumbria University. Her research focuses on green crimes, such as wildlife trafficking and animal welfare, and these crimes' intersection with organized crime and terrorism. She also conducts research around invisible crimes, including crimes of the powerful such as bio-piracy.

Contributors

Kimberly L. Barrett Sociology, Anthropology and Criminology Department, Eastern Michigan University, Ypsilanti, MI, USA

Lieselot Bisschop John Jay College of Criminal Justice, New York, USA

Ghent University, Ghent, Belgium

Lorraine Elliott Australian National University, Canberra, Australia

Judith van Erp Utrecht University, Utrecht, The Netherlands

Marieke Kluin Leiden University, Leiden, The Netherlands

Sarah Kosmicki Oklahoma State University, Stillwater, USA

Michael A. Long Department of Social Sciences & Languages, Northumbria University, Newcastle upon Tyne, UK

Michael J. Lynch Department of Criminology, University of South Florida, Tampa, FL, USA

Bill McClanahan University of Essex, Colchester, UK

Angus Nurse School of Law, Middlesex University, London, UK

Nigel South University of Essex, Colchester, UK

Toine Spapens Tilburg University, Tilburg, The Netherlands

Paul B. Stretesky Department of Social Sciences and Languages, Northumbria University, Newcastle upon Tyne, UK

Karin van Wingerde Erasmus University Rotterdam, Rotterdam, The Netherlands

Tanya Wyatt Northumbria University, Newcastle, UK

Chapter 1
The Brownness of Green Crimes and Harms

Tanya Wyatt

Criminology has awoken to the plight of the environment. The continually growing field of green criminology is testimony to such scholarship and is dedicated to uncovering green harm and crime that injures people, other species and the planet. The crimes against the environment are varied and quite extensive. White (2011) proposes these can be categorised into three groupings: white (scientific and technological concerns), green (ecological and environmental concerns) and brown (waste, pollution and toxic concerns). This book is a unique collection focussed exclusively on the latter. Hazardous waste and pollution take on numerous forms, have various causes and impact tens of thousands if not hundreds of thousands of people.

Arguably, waste in any form can become hazardous. Most materials can degrade or contaminate the environment to a degree that some form of life can sustain injury if the waste is not properly disposed of. This is evident in that simple rubbish like drink cups and plastic bags can and do become hazardous to small nonhuman animals who become trapped in this rubbish or ingests it (Ryan et al. 2009). Rubbish of this kind and other plastics are posing serious hazards to the range of marine species that come into contact with it. Plastics in water are particularly problematic because they are exposed to a limited amount of ultraviolet light that would help them break down in addition to lower temperatures that reduce the rate of decomposition (Gregory and Andrady 2003). This has contributed to the existence of enormous plastic islands or gyres in the world's oceans. These have been growing over the last four decades in particular, with the invention and widespread use of synthetic polymer plastics (Leichter 2010). As Beck (1999) has proposed in the *World Risk Society,* humans have created numerous dangers that never existed before. In this case, the plastic gyres contain significant quantities of polychlorinated biphenyls (PCBs), which probably have been incorporated into the food supply by bioaccumulating in algae and other marine animals (Leichter 2010). The pieces of plastic become persistent organic pollutants that stay in the environment for significant amounts of time (Science News 2011). Of additional concern is that this was report-

T. Wyatt (✉)
Northumbria University, Newcastle, UK
e-mail: tanya.wyatt@northumbria.ac.uk

T. Wyatt (ed.), *Hazardous Waste and Pollution,* DOI 10.1007/978-3-319-18081-6_1

ed to be a growing problem 40 years ago, when Venrick et al. (1973) documented large amounts of plastic debris in the Pacific Ocean. Of further concern still is, as Venrick and colleagues pointed out, that litter and poor disposal on land ends up in the remotest parts of the oceans and injures the environment and wildlife there. The plastic also ends up on remote landmasses. In Hawaii, for instance, thousands of miles from the nearest continent, volunteers clean up 1.5–2 tons of rubbish on the beaches of just one portion of the Big Island alone (Science News 2011).

Other waste may pose hazards on a similarly large scale. The dumping of electronic waste or e-waste from computers and appliances containing toxic materials like mercury and lead is becoming prevalent as illustrated in Chap. 3 of this book. Other hazardous waste includes spent nuclear fuel, space debris and by-products of transportation, chemical and pharmaceutical industries (see Chap. 10 regarding seagoing vessels, for example). With regards to the former, in the USA, spent nuclear fuel can be kept in cooling pools for up to 60 years at the site of a reactor. This has, since the Fukushima disaster and with ongoing terrorism concerns, been recognised to be a safety and security concern (Stewart and Stewart 2014). The alternative is to transfer it to dry casks, but the cost is seen as prohibitive and the added safety unknown. This is in lieu of what was planned, which was a single repository for most, if not all, spent nuclear fuel at Yucca Mountain in Nevada, which is tied up in a legal battle and therefore no longer an option (Stewart and Stewart 2014). The safety of long-term storage measures is unknown. This means that hazardous waste is not only a concern for those attempting to dispose of and store it but also has intergenerational implications as hundreds of years in the future people will still have to monitor the safety of spent nuclear fuel.

This is true of space debris as well, which we are beginning to feel the effects of from previous generations. For example, 22,000 pieces of debris that are over 10 cm in size are being tracked by the US Space Surveillance Network (POST 2010). This is not only to see if it falls from orbit but also to monitor if it is colliding with functional equipment, like satellites, in space. It is thought though that there is much more debris that cannot be tracked due to its location and size (POST 2010). Even small objects can pose a great risk to 'critical assets' and the health of the overall space environment (Liu 2014). Furthermore, there is no indication that space debris will decrease; in fact it is predicted that it will continue to grow unless active removal programmes are undertaken (Liu 2014).

As is evident from these illustrations, though hazardous waste is a current problem, we still must contend with hazardous waste produced decades ago. In the USA, for instance, there are 1700 so-called 'superfund' sites that are areas of contaminated land or hazardous waste storage (Voosen 2014). The name 'superfund' stems from money allocated by the US government to clean up or mitigate the risks of these toxic areas. The need for such funds was realised after the Love Canal disaster, where a residential neighbourhood had been built atop land where hazardous waste had been dumped and not declared. Residents successfully fought for compensation and relocation after the ground began seeping waste and people became ill. The number of superfund sites means that one in six Americans lives near contaminated land (Voosen 2014). As Stretesky and Hogan (1998) discovered in Florida, further concerns arise from superfund sites besides pollution. Frequently, these toxic areas are connected to environmental injustices, as ethnic minorities are more likely to live in these polluted locations. Although the US Environmental Protection Agency (EPA) has made progress in cleaning up these sites since the 1970s, there are still 95 uncontrolled sites that will possibly result in people being exposed to toxins (Voosen 2014). Hazardous waste then seems to be localised in specific spaces, often soil and water and as described, in outer space.

This is in contrast to pollution, which not only impacts soil and water but can also be seen as nearly ubiquitous in the air. Climate change is a clear example of this, where despite some continuing denial of its existence, scientists have reached a consensus that the planet is undergoing anthropogenic climate change (IPCC 2008; Stern 2007). This will not only impact people in different regions in different ways but also will impact everyone to some degree. Although most people concentrate on the alterations to their environment, be it higher sea levels, an increase or decrease in rain, and so forth, it is also predicted that climate change will 'increase strain, reduce social control, weaken social support, foster beliefs favourable to crime, contribute to traits conducive to crime, increase certain opportunities for crime and create social conflict' (Agnew 2011, p. 21). This is because climate change is linked to food and water shortages, land degradation, hurricanes and other severe weather events like droughts (Agnew 2011). These too have criminogenic consequences, so impacts of climate change will not only be on the environment but also on the social fabric of our communities.

Additionally, in localised high concentrations, air pollution, like high levels of carbon that have contributed to climate change, can have potentially devastating consequences. This is evident in China, where in some areas pollution is so high that scientists are finding that it is interfering with photosynthesis of the plants being grown as crops and in the environment (Kaiman 2014). Clearly, this possibly has far-reaching impacts, including disrupting the food supply (Kaiman 2014). The economic effects are already being felt as planes are grounded and factories are closed when pollution becomes too severe to safely operate. The human health implications are equally worrying. Air pollution levels in Beijing during February 2014 were 20 times higher than the safe levels established by the World Health Organisation. In this instance, the pollution is particularly hazardous as the particles are so small that they are capable of entering deep into the lungs and into the bloodstream of those breathing the polluted air (Kaiman 2014). China's population may suffer the consequences of this toxic air for decades to come. Though efforts such as school closures and reduced traffic on particularly smoggy days are sometimes implemented, overall the situation does not seem to be improving (Kaiman 2014). Staying indoors in China though may not reduce a person's risk of avoiding air pollution. Zhang and Smith (2007) found that nearly all rural residents and a portion of urban residents use wood, crop residue or coal for indoor household cooking and heating. In fact, indoor pollution may account for 40 % more premature deaths than outdoor pollution (Zhang and Smith 2007). Studies in China and other locations have linked the particles produced from burning wood and coal inside to increased risk of lung cancer, respiratory illnesses, reduced lung function, immune system impairment and carbon monoxide poisoning (Zhang and Smith 2007).

With much of the world struggling with contamination, it begs the exploration of who is responsible. The answer to some degree is all of us. For example, in the USA, every year 25 billion Styrofoam cups are thrown away (Carry Your Cup 2010). An additional 2.5 million plastic bottles are thrown away every hour. Even with the public's growing interest and understanding of the scope of the problem with brown and green crimes, this does not seem to alter behaviour. In the UK, plastic bag use has risen for the fourth year in a row to 8 billion, single use bags being given out in 2013 (Vaughan 2014). Around the globe in 2002, the Worldwatch Institute (2013) reports that 4–5 trillion plastic bags were manufactured. It is this type of consumption that contributes to the plastic islands in the ocean. In terms of individual accountability, interestingly, Pearce and Farrell (2011) point out that volume crime itself (such as personal assault and crimes between citizens) has a carbon footprint and preventing crime has a greening impact. So, even seemingly unrelated individual actions impact upon the health of the environment and can produce waste and pollution.

Individuals, of course, are not the only offenders, so too are companies and businesses. This happens through the products that they create (electronics, chemicals, herbicides, pharmaceuticals, and so forth) and also through the by-products that are created in manufacturing and industrial processes that take place in the diverse range of industries that support human societies. Waste and pollution can also be generated when rather than properly engaging with disposal and/or recycling protocols companies choose to cut corners to save time and money. For example, a business that fly tips instead of taking their rubbish to a landfill. Stretesky et al. (2013) argue that the treadmill of production inherent in capitalist systems embeds criminal activities like this in society, as there is always the pressure to make more money and for economic growth. This is despite the fact that the planet and its natural resources cannot sustain the level of consumption required by this form of political economy (Stretesky et al 2013).

States with their neo-liberal tendencies and lack of political will could also arguably be blamed for the scale of hazardous waste and pollution. This is connected to their limited action on climate change and their failure to take meaningful steps to mitigate the causes of global warming. The limits on carbon emissions have consistently been lowered to meet the neo-liberal agenda of growth and continued profit. States are also brown crime offenders in the actions of their militaries. For instance, on the small island of Vieques, part of the US territory of Puerto Rico, the residents are left with contamination after six decades of the US Navy testing nuclear bombs there. It is now listed as a superfund, as described above, but the US EPA projections for cleanup have shifted from 2020 (17 years after the bombing stopped) to 2029 (Stanchich 2013). Residents of the island have nearly 30 % higher cancer rates than other Puerto Ricans and are surrounded by mercury, lead and depleted uranium along with abandoned ammunitions that are in the process of being collected (Stanchich 2013). The offenders guilty of causing and disposing of hazardous waste and for polluting the environment is not limited to any one group; it is individuals, companies, governments, and so on—any person or collective.

It is surprising that with the extent of the hazardous waste and pollution that surrounds people, the impact on us is not more widely embraced by citizens and more activism is not present to demand a healthier planet. Lynch (2013) illustrates the point by comparing street crime statistics in the USA to the number of environmental violations that victimise people. In the USA, in 1 year there are approximately 25 million incidents of crime and 9 million of these are violent. Yet, there are 90 million people exposed to air pollution violations and a further 90 million people exposed to water pollution violations daily. This does not take into account the 4 million people that are living near toxic waste sites. Exposure to pollution is not normally conceptualised as violence, though arguably could be (Walters 2014). Furthermore, this is only the anthropocentric take on the victims of pollution. This in no way regards the environment itself and the other species affected by hazardous waste and pollution. So, whilst causing a significant amount of injury and suffering, criminalisation and regulation of such actions remain contentious at the same time that green efforts (grey water capture, and so forth) are prohibited. The ongoing battle in the USA to stifle actions of the EPA when they move to regulate or control carbon is evidence that the pollution and waste are not a priority and to a degree are protected to secure economic growth.

This collection of chapters, which were papers presented at the Economic and Social Research Council and Northumbria University-funded Brown Crime: Hazardous Waste and Pollution conference in October 2013, explores the dynamics of conceptualisation, control and regulation of these green crimes and those green harms outside of the criminal justice system. The collection is broken up into three parts of three chapters each—Understanding Hazardous Waste and Pollution, Environmental Justice Concerns and Corporations and Brown Crime. Part 1 goes into more specifics about particular aspects and types of brown crime. Nigel South begins in Chap. 2 by providing more detailed examples of brown crimes and sets them within the current global context of expanding neo-liberal economic policies, which relentlessly pursue profit and growth whilst minimising regulation. He questions the justice of such an approach and whether this is desirable. Further evidence for minimal regulation and its impact on hazardous waste appears in Chap. 3, where Lieselot Bisschop explores the governance of the growing flows of electronic and electrical equipment waste. She documents the challenges facing not only government but also corporate and civil society stakeholders in controlling and preventing the illegal trade and the resulting environmental harm from this hazardous waste. Lorraine Elliott then, in Chap. 4, investigates another illegal trade, that of ozone-depleting substances, a known cause of significant pollution. Her discussion highlights how the Montreal Protocol to phase out these substances unintentionally created the space for a global profitable black market.

Part 2 investigates through three different brown crimes the environmental injustices that are tied to the creation and existence of hazardous waste and pollution. Bill McClanahan analyses the binary between the access to and the pollution of water in both the Global North and Global South. He challenges the common assumption that the North is water rich and only concerned with pollution and that the South is water poor and only concerned with access. Breaking this prevailing notion reveals questions of environmental justice and human rights. Further inequalities in relation to some people's exposure to hazardous waste and pollution is brought to light in Chap. 6 where Sarah Kosmicki and Michael Long present evidence of the disparate demographics where coal and nuclear power plants are situated in the USA. They explore the differing make up of the communities that are located in proximity to these sites and investigate the causes of these differences. In Chap. 7, Michael Lynch, Michael Long, Kimberly Barrett and Paul Stretesky offer a theoretical explanation for green crimes and harms by proposing application of the treadmill of production which argues that crimes of ecological disorganisation (like brown crimes) are the result of the productive forces and mechanisms employed by capitalist economies.

Chapter 8, the first in Part 3—Corporations and Brown Crime—is Marieke Kluin's in-depth investigation into the workings of the Dutch chemical industry. She explores the nexus of violations, compliance and safety through a mixed-method approach of interviews, surveys and ethnographic observations and discovers there is little evidence of a connection. Angus Nurse then evaluates the numerous environmental harms perpetrated by oil companies in the Niger Delta. Within his discussion, he highlights the role of Corporate Environmental Responsibility (CER) in attempting to regulate companies' behaviour and comes to the conclusion that CER itself needs more regulation if pollution is actually to be prevented. Finally, Chap. 10 by Judith van Erp, Toine Spapens and Karin van Wingerde uncover the polluting behaviours of sea-going vessels that are part of the shipping and oil industries. They propose that with high costs for proper waste disposal and tremendous limitations on law enforcement to police sea-going vessels that alternative extralegal approaches must be developed to curb this significant source of pollution.

The book then is an innovative and original collection not only in terms of criminology but also in terms of the growing scholarship around green criminology. Many of these issues, like black market ozone-depleting substances and polluting sea-going vessels, to name just two, are concerns that rarely, if ever, get analysed under a criminological lens. Our aim is to continue such research and maybe to inspire others to join the effort to prevent and curb the spread of hazardous waste, pollution and toxins on a planet that is already burdened with the waste from previous generations.

References

Agnew, R. (2011). Dire Forecast: A theoretical model of the impact of climate change on crime. *Theoretical Criminology, 16*(1), 21–42.

Beck, U. (1999). *World risk society*. London: Polity Press.

Carry Your Cup. (2010). Get the facts. http://www.carryyourcup.org/get-the-facts. Accessed 18 Feb 2015.

Gregory, M., & Andrady, A. (2003). Plastics in the marine environment. In A. Andrady (Ed.), *Plastics and the environment* (pp. 379–402). New York: Wiley.

Intergovernmental Panel on Climate Change. (2008). *Climate change: Synthesis report*. Geneva: IPCC.

Kaiman, J. (2014). China's toxic air pollution resembles nuclear winter, says scientists. *The Guardian*. http://www.theguardian.com/world/2014/feb/25/china-toxic-air-pollution-nuclear-winter-scientists. Accessed 11 Jan 2015.

Leichter, J. (2010). Investigating the accumulation of plastic debris in the North Pacific Gyre. In K. Omori, X. Guo, N. Yoshie, N. Fujii, I. C. Handoh, A. Isobe, & S. Tanabe (Eds.), *Interdisciplinary studies on environmental chemistry—marine environmental modeling & analysis* (pp. 251–259). Tokyo: Terrapub.

Liu, J. (2014). Modeling the large and small orbital debris populations for environment remediation. National Aeronautics and Space Administration, 3rd European Workshop on Space debris modeling and environment remediation, Paris, France. http://ntrs.nasa.gov/archive/nasa/casi.ntrs.nasa.gov/20140006500.pdf. Accessed 17 Feb 2015.

Lynch, M. (2013). Reflections of green criminology and its boundaries: Comparing environmental victimization and considering crime from an eco-city perspective. In N. South & A. Brisman (Eds.), *The Routledge international handbook of green criminology*. (pp. 43–57). London: Routledge.

Parliamentary Office of Science and Technology, Space Debris (2010). Postnote. Mar 2010 No 355, 1–4.

Peace, K., & Farrell, G. (2011). Climate change and crime. *European Journal of Criminal Policy and Research*. doi:10.1007/s10610-011-9143-1.

Ryan, P., Moore, C., van Franeker, J., & Moloney, C. (2009). Monitoring the abundance of plastic debris in the marine environment. *Philosophical Transactions of the Royal Society B, 364*, 1999–2012.

Science News. (2011). Swirling seas of plastic trash. *Science News for Kids*. http://www.sciencenewsforkids.org/2011/06/swirling-seas-of-plastic-trash/. Accessed 13 Jan 2015.

Stanchich, M. (2013). Ten years after ousting US Navy, Vieques confronts contamination. *The Huffington Post*. http://www.huffingtonpost.com/maritza-stanchich-phd/ten-years-after-ousting-u_b_3243449.html. Accessed 18 Feb 2015.

Stern, N. (2007). *The economics of climate change*. Cambridge: Cambridge University Press.

Stewart, R., & Stewart, J. (2014). Solving the spent nuclear fuel impasse. *New York University Environmental Law Journal, 21*, 1–142.

Stretesky, P., & Hogan, M. (1998). Environmental justice: An analysis of superfund sites in Florida. *Social Problems, 45*(2), 268–287.

Stretesky, P., Long, M., & Lynch, M. (2013). *The treadmill of crime: Political economy and green criminology* (*new directions in critical criminology*). London: Routledge.

Vaughan, A. (2014). Plastic bag use rises for fourth year. *The Guardian*. http://www.theguardian.com/environment/2014/jul/15/plastic-bag-use-rises-for-fourth-year. Accessed 18 Feb 2015.

Venrick, E. L., Backman, T. W., Bartram, W. C., Platt, C. J., Thornhill, M. S., & Yates, R. E. (1973). Man-made objects on the surface of the central North Pacific Ocean. *Nature, 241*, 271.

Voosen, P. (2014). Superfund sites. *National Geographic*. http://ngm.nationalgeographic.com/2014/12/superfund/voosen-text. Accessed 11 Jan 2015.

Walters, R. (2014). Air pollution and invisible violence. In P. Davies, P. Francis, & T. Wyatt (Eds.), *Invisible crimes and social harms* (pp. 142–160). Basingstoke: Palgrave Macmillan.

White, R. (2011). *Transnational environmental crime: Toward an eco-global criminology*. London: Routledge.

Worldwatch Institute. (2013). New bans on plastic bags may help protect marine life. http://www.worldwatch.org/node/5565. Accessed 18 Feb 2015.

Zhang, J., & Smith, K. (2007). Household air pollution from coal and biomass fuels in China: Measurements, health impacts, and interventions. *Environmental Health Perspectives, 115*(6), 848–855.

Tanya Wyatt is a Principal Lecturer in Criminology at Northumbria University.

Part I
Understanding Hazardous Waste and Pollution

Chapter 2
Green Criminology and Brown Crime: Despoliation, Disposal and De-manufacturing in Global Resource Industries

Nigel South

This chapter adopts a green criminological orientation to consideration of several examples of what has been termed 'brown crime'. In the following, these terms are defined and then the context of current neo-liberal economics is described. In the rest of the chapter, a series of cases are presented to illustrate the consequences and challenges arising from our routine use of our environments, landscapes and planet as a wellspring of economic goods and as a convenient dumping ground for 'economic bads'.

Green Criminology and Brown Crime

'Green criminology' is an umbrella term used to cover and capture the study of ecological or environmental crime or harm, and related matters of speciesism and environmental (in)justice. It provides a perspective and loose framework of theories and methods to apply to the investigation of harms, offences and crimes related to the environment, different species and the planet (see, e.g. Beirne and South 2007; Sollund 2008; South 2014; South and Brisman 2013; White, 2008, 2010). Importantly, it is 'open' to inter- and multi-disciplinary engagement. For present purposes, it is helpful to quote Walters' (2010, p. 181) description of 'eco-crime' as covering 'The contamination of drinking water, the degradation of soil and the pollution of air and land (all of which) expose people (usually those in poor and developing countries) to substantial health risks'. As Walters points out, such acts are frequently 'linked to the poverty and social dislocation, as well as the mental and physical debilitation, of people who are victims of corporations and states that deliberately violate environmental agreements'.

Ruggiero and South (2010) have described the phenomenon of 'dirty collar crime' whereby legitimate businesses are involved in semi-legal or wholly illegal

N. South (✉)
University of Essex, Colchester, UK
e-mail: soutn@essex.ac.uk

T. Wyatt (ed.), *Hazardous Waste and Pollution,* DOI 10.1007/978-3-319-18081-6_2

waste disposal operations that can contribute to the creation of environmental and public health harms or disasters, such as the Naples garbage crisis of 2008 and 2010, yet still make significant financial profits and improve their standing through networks of influence and enterprise. The waste dumping at the heart of this particular case is an example of what might be called 'brown crime'. White (2008, pp. 98–99) defines this term as part of a threefold set of 'brown', 'green' and 'white' categories of environmental issues in which 'brown' refers to urban life and related pollution (e.g. air pollution, disposal of toxic/hazardous waste, oil spills, pesticides); 'green' refers to conservation and wilderness issues (e.g. acid rain, biodiversity loss, habitat destruction); and 'white' refers to the impact of new technologies and laboratory practices (e.g. animal testing, cloning, environmentally-related communicable diseases, genetically modified organisms).

The Context of Neo-liberalism

As McCarthy and Prudham (2004, p. 275) observe, 'connections between neo-liberalism, environmental change and environmental politics remain under-explored in critical scholarship' and yet are 'deeply if not inextricably interwoven'. One initial obstacle to such exploration is defining the nature of neo-liberalism which is a discursive and dispersed 'assemblage' of commitments, representations and practices (McCarthy and Prudham 2004). Adopting Jessop's (2002, p. 461) straightforward (although, as he would emphasise, only partial) definition, neo-liberalism can be seen as promoting 'market-led economic and social restructuring' with outcomes for the public sector that involve 'privatization, liberalization and imposition of commercial criteria in the residual state sector', while in the private sector, 'deregulation is backed by a new juridico-political framework that offers passive support for market solutions'.

In the realm of economic ideas, the current influence of neo-liberalism has promoted a set of philosophies and policies that argue 'no-limits growth' is desirable and sustainable. In this view, the enhancement of efficiency and productivity requires the minimisation of regulatory measures and prescriptive arrangements designed, for example to reduce risks to health and safety or negative environmental impacts. In the neo-liberal account, regulation is largely unnecessary as environmental harm is accidental, unintentional and external. Economic growth necessitates the removal of environmental safeguards that act as obstacles. When damaging side effects of growth do occur, this is regrettable but justifiable because these are also products of necessity (Ruggiero and South 2013a, b). Many economic activities impose significant costs on the well-being of humans, other species and ecosystems generally but these are simply seen as the 'costs of doing business'. From the neo-liberal point of view, the planet is a resource available to those who are most capable of exploiting it, and human skills of enterprise, survival and exploitation are seen as evolution's evidence that the restraints and resistance of nature can and should be overcome. This is a contemporary rationale and justification for the ever-increasing plundering and pollution of the planet but also traces a philosophical line of supportive argument to the father of classical liberalism, John Locke (1632–1704). As Broswimmer (2002, p. 57) notes, Locke argued that 'nature was given to the "industrious and rational"' and he viewed 'the whole of nature as a mere resource for commercial exploitation, arguing that "land that is left wholly to nature is called, as indeed it is, waste"'. Broswimmer (2002, p. 57) argues that the legacy of such thinking is that 'The sanctification of private property in the hands of liberal thinkers has played a crucial part in the emergence of global capitalism. At its very core, the prevailing capitalist ethos and liberal world view of the modern industrial era remained expansionary and imperial involving a calculated form of indifference to the social and ecological order'. McCarthy and Prudham (2004, p. 277) argue it is therefore:

> salient to remind ourselves of how centrally and explicitly liberal thinking itself turned on restructuring social relations to nature. This process is most infamously associated with enclosing commons to facilitate the development of increasingly capitalist, export oriented farming operations … . Such reconfigurations of property relationships amounted to "freeing" up nature, i.e. detaching it from complex social constraints and placing it under the auspices of the self-regulating market …, whilst jump-starting capitalism through primitive accumulation, or what Harvey (2003) has recently termed, "accumulation by dispossession". In turn, the emergence of new social relations to nature factored centrally in classical liberal ideological, discursive, and material practices, all of which have parallels in and influences on neo-liberalism.

This chapter presents examples of 'brown crimes' or harms that arise from modern excesses of neo-liberalism, as it demands and leads to processes and problems of despoliation, disposal and de-manufacturing that are central to global resource industries: oil pollution; dispersal of residues of dangerous and radioactive waste; chemical warfare and its legacies; asbestos dust and effluvia; and harms and hazards related to the global recycling economy. It concludes with some observations on the power of offenders to disregard or disempower regulatory governance, and some signposts to future challenges and responses.

Oil Crimes and Spoiling of the Environment in Nigeria

This first section describes problems related to gas flares, oil spills and pollution, poverty and abuse of rights which can be described as 'environmental racism'. To explain the problem, gas flares are the result of burning unwanted 'associated gas' that is produced during the process of pumping oil from the ground. This flaring produces toxins in the atmosphere that rain into the swamps, creeks and forests, acidifying the rain and polluting the soil. According to Howden (2010), 'Medical studies have shown the gas burners contribute to an average life expectancy in the Delta region of 43 years. The area also has Nigeria's highest infant mortality rate'. Flaring has been illegal in Nigeria since 1984 and three deadlines to cease the practice have been missed so far. Instead, new facilities and sites of flaring have been established as, for example in 2010 in the Niger Delta. Howden quotes Alagoa Morris, an investigator with a local Environmental Rights Action group, who describes this continuing pollution of air, water and land as 'environmental racism' and who says 'What we are asking is that oil companies should have to meet the same standards in Nigeria that they do operating in their own countries'. Morris, says Howden, 'regularly risks arrest to monitor activities at the heavily guarded oil and gas installations'. Considerable attention has been drawn to this ongoing problem by the NGOs and journalists, but oil companies have failed to act to cease or effectively remedy damage done, instead resorting to techniques of neutralisation of the problem, acknowledging its past significance but minimising its current impact. So, for example as Shell Global (2013) stated: 'Oil fields produce a mix of oil, water and natural gas. In the past it was standard industry practice to burn off, or flare, the gas if there was no market for it. But this was a waste of a valuable resource and produces carbon dioxide'. In a fine example of corporate under-statement, Shell Global continued, 'It can also cause disturbance to local communities that have often grown up around the flare pits'. The acknowledgment of the sizeable scale of the problem comes in a separate statement that seeks approval for how well they are doing in reducing it: 'SPDC (Shell Petroleum Development Company of Nigeria) reduced flaring volume from its facilities by about 75 % between 2003 and 2012 and flaring intensity (the amount of gas flared per barrel of oil produced) by around 60 % over the same period' (Shell companies in Nigeria 2014). Continuing work on reduction of flaring is being supported and Shell asserts that 'When all this work is completed it will cover over 90 % of the SPDC's production potential'. (Shell Global 2013). Overall, this is reminiscent of many corporate narratives of neutralisation—'yes, it was a problem but we are working on it and when we've finished it won't seem such a problem after all'. But the situation also reflects three other issues. First, the possibility of alternative but unacknowledged explanations for the reductions achieved. So, as Howden (2010) notes, although Nigerian officials have claimed 'record reductions in the amount of gas flared, independent oil and gas experts believe flaring is, in fact, reaching historic highs. Many observers attribute ... [a] much-trumpeted reduction' in 2009 to the effects of militant action against oil companies in the Niger Delta which led to the halving of oil production. Second, denial of the damage already done. Howden reports that what has happened in the Niger Delta is, according to independent oil and gas expert, Chris Cragg, a 'continuing economic, political and environmental disaster [and] one of the largest single pointless emissions of the greenhouse gas on the planet, with obvious implications for climate change that will not only affect Nigeria, but also the rest of the world'. And finally, of course, there should be a genuine motivation to now do something about this 'pointless' flow of emissions—even if not for environmental then at least for profit-motivated reasons. If the amount of wasted, flared natural gas were captured and cycled though a modern power station the volume produced 'could fuel about a quarter of Britain's power needs' being 'equivalent to more than one third of the natural gas produced in the UK's North Sea oil and gas fields' and enough to 'meet the entire energy requirements of German industry'. So the gas could be profitably productive instead of generating pollution that 'has been measured at up to 50 million tonnes of carbon dioxide, with unknown quantities of the far more damaging greenhouse gas: methane' (Howden 2010). Neither the Nigerian government nor the oil companies have been sufficiently motivated by the prospect of this possible source of profit, but have been de-motivated by tough investment and infrastructure challenges and so simply continue to take the easy route to profitable extraction while leaving the pollution costs to be paid by the local people and the environment.

In 2011, the UN Environment Programme (UNEP) reported on over 50 years of pollution as a result of oil production in Ogoniland, in the south-east of the Niger Delta region, and pointed out that the Nigerian Government has failed to follow or enforce its own laws and requirements. This is an example of the kind of 'secondary green crimes and harms' described elsewhere (Carrabine et al. 2014, p. 391, pp. 404–405) as 'the violation of rules that attempt to regulate environmental harm and to respond to disaster' and that can include 'major and minor practices whereby states violate their own regulations (either by commission or omission) and in so doing contribute to environmental harms'. As Vidal (2011) reports, Audrey Gaughran of Amnesty International has argued that 'Oil companies have been exploiting Nigeria's weak regulatory system for too long'.

In fact, regulatory systems allow oil companies to provide key measures of the extent of oil spills and this (unsurprisingly) may often yield an under-estimate of the true extent of spill and damage (Shrope 2013). And this, of course, relates to cases that are known and reported. In many cases, oil spills, whether deliberate or accidental, occur with no known identified source and if they occur at sea then unless the effects are noted by others at sea or from the air, the original location of the pollution will be unknown although the results of spillage such as damaged and dead birds and fish, as well as pollution arriving at the shore, will eventually indicate an incident (see, e.g. Morris 2013, p. 4). There are spills, leaks and ecological damage occurring on a highly regular basis across the world—the majority never attracting the kind of publicity attached to the BP Deepwater Horizon/Gulf of Mexico. In fact, writing one month after the Gulf of Mexico oil spill, Vidal (2010) pointed out that the Niger Delta suffers more pollution *every* year than had been caused at that point by the Deepwater Horizon disaster and that the contrast in reactions was striking: 'the Niger delta ... is the world capital of oil pollution. Life expectancy in its rural communities, half of which have no access to clean water, has fallen to little more than 40 years over the past two generations. Locals blame the oil that pollutes their land and can scarcely believe the contrast with the steps taken by BP and the US Government to try to stop the Gulf oil leak and to protect the Louisiana shoreline from pollution'(see also Pegg and Zabbey 2013). Occasionally, compensation is paid as agreed in January 2015 when Shell announced it would pay out £ 55m to 15,600 Ogoni locals whose farming and fishing were devastated by oil spills in 2008 and 2009.The money will be split between the individuals and the community, but as Vidal (2015) reports, the settlement 'avoids Shell having to defend a potentially embarrassing London high court case which was due to start shortly'. And as Vidal also reports, a separate development reflects welcome remediation, but also the possible persistence of the tendency, noted above, to under-estimate the depth and breadth of pollution problems. In this case,

> the company's Nigerian subsidiary Shell Petroleum Development Company of Nigeria (SPDC) said it expected to start to clean up its pollution in the Bodo fishing grounds and swamps "within months". Shell had initially estimated that around 4,000 barrels of oil were spilt in ... two events, but oil experts calculated from film footage that it could have been 60 times as much. According to Amnesty International, Shell had intentionally underestimated the spills in an attempt to minimise compensation payments. This was denied by Shell.

In Nigeria, state military forces and private mercenary companies have been involved in operations to secure oil facilities thereby protecting the interests of those who benefit from oil wealth and assert a right to pollute at the expense of the living standards and human rights of those who live in these oil producing areas. Importantly, while some oil wealth remains in the country, distributed among a small elite much, of course, also leaves the country and benefits the corporate owners of such national subsidiaries.

Around the world, whether in developing or developed nations, communities live on dangerously polluted land and suffer the daily consequences and long-term legacies of damage to environments and health caused by the industrialisation of hazard. From the people of Jharia, India who live on top of an open-cast mine that is permanently burning above ground (Global Journal 2012) to workers and residents affected by the depleted uranium contamination left behind after a factory closure in Colonie, in upstate New York (Rose 2007), to the involvement of organised crime groups in disposing of hazardous wastes by mixing it with materials 'to make bricks or resurface roads, and use raw materials to make fertiliser which subsequently transfers chrome, cadmium, lead and nickel up the food chain' (Liddick 2010, p. 139). In the next section two cases of such 'routine' disposal of waste are outlined, one with devastating consequences for members of the community, and the other representing a strategy of 'denial of danger' offered by simply dumping waste at sea.

Disposal and Dispersal of Dangerous Waste

The Corby Case

In the UK, from 1940s onward, Corby in Northamptonshire was the home of a massive steel works which over the 46-year history of operational life 'produced a dizzying array of dangerous waste—nickel, chromium, zinc, arsenic, boron and cadmium' (Gordon, 2009). At the end of the life of the site, when British Steel closed it in 1980, the local authority took control and was faced with the challenge of disposing of the waste. This they proceeded to do, 'in the back of open lorries, sludge spilling onto the public roads of the town', with one local remembering 'the smell and the metallic taste of it, and how if you drove behind one of the lorries, your car always ended up covered in a light film' (Gordon 2009). Reporting as the High Court heard a group litigation case against Corby Borough Council at the end of July 2009, Gordon (31 July, 2009) records that the court heard how: waste was dumped all over Corby by staff that Mr Justice Akenhead described as being 'unqualified and insufficiently experienced' and how a waste management expert who saw how the materials were disposed of, was said to have been 'appalled'. At the time that the land was being 'reclaimed', an auditor described the operation as 'naïve, cavalier and incompetent'. After a 10 year battle, the Judge ruled that Corby Borough Council had been negligent and that the dumping of toxic material may have caused birth defects in children. This was a case described by lawyers acting for the affected families as: 'the biggest child poisoning case since Thalidomide' (Gammell 29 July 2009). Nearly 1 year later, in April 2010, Corby council withdrew its legal challenge and reached an agreement to pay compensation to the affected children albeit without accepting liability in the case.

Radioactive Waste Dumping in the English Channel

Walters (2007, p. 188 and passim) has drawn attention to the eco-crime links between nuclear industry activities and disposal of radioactive waste, noting the variety of 'risks associated with commercial enterprises in research, power production, telecommunications, medicine and pharmaceuticals as well as state activities in military defence and war' that 'all utilise varying degrees of radioactive substances that produce waste'. Although radioactive waste may be recycled in some forms and can be exported legally or illegally, and while it has a high value for some purposes, it is a hazardously difficult commodity to manage, posing complex logistical and expensive challenges for proper storage. One way to avoid these challenges has been to simply bury it or dump it at sea (Ringius 2001; Parmentier 1999). In the past, the Russian navy has disposed of submarine reactors and nuclear waste in the Barents and Kara seas, while in 2000 Greenpeace exposed a UK policy operating between 1950 and 1963 that meant containers of nuclear waste were simply dumped near the Channel Islands. Approximately 28,500 corroding containers were discovered, with this being just one of many dumpsites used before the global banning of the practice in 1993 (Greenpeace 2000).In this respect, such disposal is a stark example of the disregard of hazard and the denial of consequences: of all the materials humanity may choose to dispose of in the sea, radioactive waste and functional but decommissioned reactors must be high on a list of indicators of 'lack of care'.

In this latter case, the dumping programme was an authorised way of managing an industrially produced contaminant. The next two examples show, first, how dangerous uses can be deliberately developed from promising research with a very different starting point. In this case, the chemical regulation of plant growth, originally stimulated by experiments to address gaps between population growth and agricultural output (Zierler 2011, pp. 35–42). And, second, how identified dangers have long been associated with the use of what was seen as a 'miracle material' but were ignored or contested.

Chemical Hazards, Toxicity and Legacies

Thousands of new chemicals are introduced into products reaching the market every year. These are the outputs of what Zierler (2001, p. 47) calls a staggeringly influential industry which has adopted an aim of being able to offer 'control of an unruly natural world through chemicals'.

However, it is difficult to estimate the long-term effects of having so much chemical presence and exposure in our environments and research on debilitating or deadly effects of such substances is hampered not only by lack of funding for studies that may be critical and lead to controls but also because violators are often successful in presenting violations of laws and regulations as 'accidents' thereby eroding accumulation of evidence that such research is needed (Pearce and Tombs 1998).

Toxic Chemicals: The Case of Dow and the Legacy of Agent Orange

Dow Chemical was founded in 1947 and in 2001 merged with Union Carbide (a company not without its own history of catastrophic accident). The company is a global operator, producing chemicals and plastics for a variety of different markets and, as Katz (2010) points out, can call on more financial and legal resources than any agencies charged with the task of regulation. Katz provides a helpful history of Dow Chemical's involvement in the production of herbicides that then found particular utility as a military weapon (Zierler 2001, pp. 46–47). In the 1960s, Dow developed these substances for use as the main ingredients in Agent Orange, the toxic defoliant used during the Vietnam War between 1961 and 1971 to destroy food crops and jungle cover. At least 20 million gallons (as well as other herbicides) were sprayed over 'enemy territory' in Vietnam, Laos and Cambodia and affected 8600 square miles of jungle and cropland, (The Week 2012; there is a parallel history of the development and use of napalm with 388,000 tons dropped on Indochina between 1963 and 1973: see Neer 2013).

The effects and legacies were a crime against humanity and against the environment. In 1971, results of tests conducted by the US National Institutes of Health showed chemical contamination from Agent Orange caused birth defects in laboratory animals and in response the USA ceased use. However, by then the chemicals were responsible for both immediate impacts as well as a lasting legacy with 100,000s of civilians and soldiers (Vietnamese and American) exposed, and waterways, soil and the food chain compromised and affected. Leukemia and blood disorders, heart disease, children with birth defects (e.g. spina bifida, limb and bone defects) were all reported health consequences (The Week 2012).And the legacy is still significant today, as acknowledged by the USA which has committed to spending $ 44 million between 2012 and 2016 'to remove dioxin residues around the former US airbase in Danang, where most Agent Orange barrels were stored. Some 2.5 million ft^3 of soil and sediment around the airport will be dug up and heated to very high temperatures, breaking down the toxic compounds. These are "the first steps to bury the legacies of our past", said US Ambassador David Shear'.

Asbestos Processing and Pollution

The mining, chemical and industrial processing of asbestos provides a case of what was once seen as a 'miracle material' or 'magic mineral' (Bowley 1960; Tweedale 2000), widely used since the days of the early Greek and Roman civilisations. And yet asbestos presents dangers at all stages of its extraction and preparation and this has been noted for centuries (Mesothelioma Center 2014). In relation to primary extraction and processing, it is potentially lethally harmful to workers and damaging to entire local communities as dust is distributed beyond sites of production and drifts into the surrounding environment, where dust and fibres are inhaled and also settle on land and water. A more formal recognition of the potential health hazard posed by asbestos was provided in Britain by the Factory and Workshop Inspectorate officers as long ago as 1898 (Tweedale 2000) but the properties and versatility of asbestos have undoubtedly explained its long use and the denial of problems associated with its production and the deterioration of the material which loosens the fibres. During the 1990s, the European Environmental Agency (EEA) estimated that during the course of the twentieth century, up to that point, around 4 million people had died in Europe from asbestos-related illnesses. In 1999, the use of asbestos was banned by the European Union, 101 years after official observation of its dangerousness. Yet the legacy has remained, as illustrated by the prosecution of the managers of the Eternit Company between 2009 and 2012. Eternit opened its first asbestos production plant in Italy, the biggest in Europe, in 1907 in Piemonte and although it had been scientifically shown and known since 1962 that asbestos dust causes asbestosis and malignant diseases, the company failed to take precautions to safeguard workers or communities (BBC News 2012). Eternit is not the only asbestos company with a poor health and safety record but is a rare exception in facing well publicised, serious and successful prosecution (although see Tweedale (2000) on the Turner and Newell litigation in England).

The cases described so far are largely problems of production, related waste disposal and industrial hazard that have been occurring across a globalising world for more than a century (Sellers and Melling 2011). However, the massive boom in production and consumption of electronic goods of the past few decades has intensified existing trends while creating new problems, markets and industries based on 'de-manufacturing'.

De-manufacturing, Dumping and Disposal

Dumping, disposal and dispersal of waste and the unwanted are familiar practices and problems. But a major development of the last few decades is not simply removing and relocating waste from the developed world to the developing world to dump there as worthless disposables, but now re-locating it as *resource-rich disposables* to be de-manufactured and recycled. Recycling is obviously 'good' but 'de-manufacturing' means that paradoxical problems arise. The recycling of the waste produced through consumption by the wealthy has consequences of concern for the labouring and scavenging poor of China and India and these consequences include both damage to the environment and damage to health and life (human and non-human).

Most particularly the USA, but additionally all developed nations, now consumes an astronomical amount of electronic goods which are manufactured and sold in a market that is premised on relentless replacement of the old by the new. Products may become obsolete by virtue of ceasing to function effectively or be desired affectively. Devices are sold on the understanding they will soon be redundant, and designed to be disposable. However, they are still valuable for their content. In principle, recycling of such materials is environmentally good but in practice a new 're-cycling for profit' 'boomerang industry' has now developed which means that waste electronic items may be exported to China where they are de-manufactured, with various parts (such as rare earths, precious metals) then reused in new electronic goods which are shipped back to advanced markets. In China, towns, factories and scrap-yards specialising in servicing this global market now store, sort and process imported items, employing a formal workforce and also creating a shadow scavenger workforce in an informal economy based upon picking over the waste of what is leftover. All are exploited in a low-paid and dangerously unhealthy industry. While the recycling and reuse of valuable resources makes sense, what is less defensible are the conditions under which the 're-cycling' occurs or the need for such a boomerang market at all. These electronic goods do not need to be disposed of so rapidly, new ones do not need to be produced and marketed with such urgency and intensity, and the environmentally good strategy of recycling should not be a cloak for an example of bad multinational exploitation in the process of which land and air are polluted by use of hazardous chemicals, and the effects on the health of workers and communities include serious diseases, some with cancer links. Huo et al. (2007) reported on hazardous chemicals released from e-wastes through disposal or recycling processes and noted that 'Past studies have reported soaring levels of toxic heavy metals and organic contaminants in samples of dust, soil, river sediment, surface water, and groundwater of Guiyu Residents' and that this has led to a 'high incidence of skin damage, headaches, vertigo, nausea, chronic gastritis and gastric and duodenal ulcers, all of which may be caused by the primitive recycling processing of e-waste'. Lead is widely used in this process and leads to a variety of health hazards, entering 'biological systems via food, water, air and soil' with children being 'particularly vulnerable to lead poisoning...' (see also Chen et al. 2011).

Discussion

'Toxic tragedies' (Cass 1996, pp. 110–112) are commonplace but difficult to prosecute due to problems of gathering evidence that ties commercial operations to specific illegal offences, cases of corruption and strong industry 'profit-at-all-costs' motivations. These cases reflect 'institutionalised insensitivity to right and wrong' (Simon 2000, p. 635) and represent profound and damaging forms of environmental injustice. Economic and environmental regulations are viewed by the neo-liberal lobby as overly costly and anti-growth. One alternative and critical view might agree that regulation should indeed be 'anti-growth' because economic development that causes environmental degradation makes a *negative* contribution to the creation of wealth and reducing growth in a managed way is desirable (Ruggiero and South 2013a) and could help shape a 'green economy'. This would be an alternative to a system that allows companies to take large public subsidies to produce profits yet leave behind eco-wreckage harming public health and suppressing ecosystem vitality which then needs further public funding to mitigate and remedy.

Regulation and compliance systems are often defended as being more effective than critics claim, but as Kostelnik (2012, pp. 148–149) summarises of regulation relating to white collar financial offending (an argument transferable to the world of environmental offending), 'There are several problems' with the standard regulatory approach:

> For one, many of the individuals in positions of power in regulatory agencies are former corporate executives. ... The presence of individuals with close ties to corporations (and presumably their interests, at times) creates an economy of crony capitalism where conflicts of interest between the governed (the corporations) and the governing (the regulatory agencies) are commonplace, often at the expense of stockholders. Second, corporations are prone to 'creative compliance', that is, manipulating regulations to their advantage (against the intent of the regulations) while remaining technically compliant. Finally, political lobbying can play a large role in amending regulations to corporations' advantage. Together, these problems have created a regulatory framework of instability where regulations are constantly being amended and changed, leading to high transaction costs of ensuring compliance and uncertainty over what corporate behaviour is morally and legally unacceptable.

Of course, national and also multi-lateral environmental regulation agreements *do* exist (the latter include, e.g. the 1989 Basel Convention on the control of Transboundary Movements of Hazardous Wastes and their disposal; the Stockholm Convention on persistent organic pollutants, in force for signatories from 2004, plus various EU laws). But these tend to be weak in implementation and impact. States and corporations may simply ignore these, while companies can adopt strategies and mount campaigns that influence and (re-)shape law. For example, big business often makes successful calls for exemption or exceptional leniency with regard to environmental regulation and argues that such intrusion is authoritarian and misunderstands the reality of business needs. An illuminating case study is provided by the actions of companies in the tobacco industry and their efforts to temper European legislation, as Smith et al. (2010) report based on their examination of internal documents from British American Tobacco (BAT) that were disclosed as a result of litigation in the USA, as well as review of other relevant literature and interviews with key informants. This evidence and analysis shows that from 1995, BAT worked with other businesses to promote European regulatory reforms that would be favourable to large corporations, in particular, the establishment of a business-orientated form of Impact Assessment. A lobbying campaign, led by BAT, but also involving a network of other companies, ensured binding changes to the Treaty of Amsterdam that require policymakers to minimise legislative burdens on businesses. This thereby shapes all future EU policy decisions and increases the likelihood that policies will benefit corporations rather than citizens. The authors also show that other business sectors such as the chemical industry—a frequent pollution offender—have since used Impact Assessments to 'delay and weaken EU regulation on the Registration, Evaluation, Authorisation and Restriction of Chemical Substances (REACH)'. (Smith et al. 2010, p. 9, Box 2).As Coll (2012) shows in his study of the 'private empire' of Exxon-Mobil oil, while the interests of corporations and states may frequently align, major global corporations will also position themselves 'above' states, whether in terms of investment decisions (choosing to invest where it is cheaper— in developing nations rather than home states) or in terms of ignoring or challenging international human rights laws or expectations. In other cases, states may not be the parties responsible for particular harms but may (for various reasons) be less than diligent in pursuit of justice when faced with pressure or financial deals (or both) from those that are responsible. For example, in the Trafigura case, following the dumping of toxic waste in Abidjan, Cote D'Ivoire and a deal that the Government subsequently made, the anti-toxic waste campaigner Helen Pervier wrote that 'The ease with which international environmental laws are broken and questionable deals exchanged for real justice, painfully highlights yet again, that the international community creates laws but simply lacks political will to implement and enforce them'. ('Greenpeace condemns the Trafigura-Cote d'Ivoire deal as travesty of justice, Greenpeace Press Release, 14 February 2007' quoted in Fagbohun 2007, p. 851).

One response might be a powerful international legal instrument such as an international crime of Ecocide that could be the basis for powerful preventative measures. According to this proposal, those in senior positions of responsibility in corporate bodies would be at risk of prosecution if they were responsible for taking decisions that lead to, support or finance mass damage and destruction. So instead of 'the polluter pays' (if caught), the new governing principle becomes 'the polluter does not pollute' and the protection of interests shifts from those few who have ownership to the many who are at risk of suffering (Higgins et al. 2013). Such a law could be supported by the implementation and operationalisation of environmental courts and the establishment of an International Environmental Court. Such courts could bring focus and expertise to bear on complex and technical matters that are often unfamiliar when introduced and processed through traditional courts (Walters and Westerhuis 2013; White 2013). Courts or other environmental public forums, promoted via other new conceptions of environmentally sensitive and sensible ways of living (e.g eco-cities: Lynch 2013), would provide and allow for greater participation of the public in matters and decisions affecting environmental regulation. This process might draw upon the notion of 'environmental due process' which is 'reflected in a number of international and regional human rights instruments, including the Universal Declaration of Human Rights ..., the International Covenant on Civil and Political Rights ..., the African Charter on Human and People's Rights ..., and the American Convention on Human Rights ..., as well as in international, regional, and national documents specific to the environment'. (Hunter et al. 2002, p. 1312; quoted in Brisman 2013).

Public understanding of the global challenges facing humanity and the planet is urgent. Key questions are: how to address energy and resource deficits and address the problem of sustainability; and, how to cope with the unwanted discards of consumer society and mounting waste? The West continues to consume at a ferocious rate and developing nations are rapidly generating their own markets and demand, in turn constantly extracting more resources from the planet. Broswimmer (2002,

103) summarises this starkly, observing that humans 'sequester an astonishing 40% of the entire terrestrial primary production of the Earth for our selfish use. Capturing such an enormous fraction of the Earth's natural productive capacity comes at a huge cost in terms of loss of natural habitat or reduction in the viability or outright mass extinction of species'.

This is a warning about risk and cost that economic systems should be able to understand and yet in the midst of a global economic downturn, remedies are viewed almost entirely in terms of *pushing up* production and *increasing* the rate of consumption in pursuit of the neo-liberal principles that enterprise and exploitation are rational and desirable. Instead it should be seen as more logical and valuable for the long term survival of the planet to turn to the question of how to *decrease* consumption and *slow down* growth. This would follow from the second key recommendation of the 2012 Royal Society Report on *People and the Planet* which argued that 'The most developed and the emerging economies must stabilise and then reduce material consumption levels through: dramatic improvements in resource use efficiency, including: reducing waste; investment in sustainable resources, technologies and infrastructures; and systematically decoupling economic activity from environmental impact'. Without such urgent action then as Broswimmer (2002, p. 103; see also South 2010) concludes, the 'synergism and combined input of contemporary military, demographic, and socioeconomic depredations suggest that the juggernaut of late modernity has entered an increasingly ecocidal phase'.

References

Beirne, P., & South, N. (2007). *Issues in green criminology*. Cullompton: Willan.

BBC News. (2012). Two convicted in asbestos-linked deaths trial in Italy. http://www.bbc.co.uk/news/world-europe-17016110. Accessed 6 June 2015.

Bowley, M. (1960). *Innovations in building materials*. London: Duckworth.

Brisman, A. (2013). The violence of silence: Some reflections on access to information, public participation in decision-making, and access to justice in matters concerning the environment. *Crime, Law and Social Change, 59*(3), 291–303.

Broswimmer, F. (2002). *Ecocide: A short history of the mass extinction of species*, London: Pluto Press.

Carrabine, E., Cox, P., Fussey, P., Hobbs, D., South, N., Thiel, D., & Turton, J. (2014). *Criminology: A sociological introduction* (3rd ed.). London: Routledge.

Cass, V. (1996). Toxic tragedy: Illegal hazardous waste dumping in Mexico. In S. Edwards, T. Edwards, & C. Fields (Eds.), *Environmental crime and criminality: Theoretical and practical issues*. New York: Garland.

Chen, A., Dietrich, K., Huo, X., & Ho, S. (2011). Developmental neurotoxicants in e-waste: An emerging health concern. *Environmental Health Perspectives, 119*(4), 431–438.

Coll, S. (2012). *Private empire: Exxon mobil and American power.* London: Penguin.

Fagbohun, O. A. (2007). Regulation of transboundary shipments of hazardous waste: A case study of the dumping of toxic waste in Abidjan, Cote D'Ivoire. *Hong Kong Law Journal, 37,* 831–858.

Gammell, C. (2009). Corby birth defects: Worst child poisoning case since Thalidomide', 29 July, *Telegraph.co.uk*. http://www.telegraph.co.uk/health/healthnews/5935220/Corby-birth-defects-worst-child-poisoning-case-since-thalidomide.html. Accessed 13 Jan 2015.

Global Journal. (9 July 2012). India: The fiery coalfields of Jharia. *The Global Journal*. http://theglobaljournal.net/group/012/article/772/. Accessed 13 Jan 2015.

Gordon, B. (2009). Corby's toxic waste: The families speak out, 31 July, *Telegraph.co.uk*. http://www.telegraph.co.uk/news/uknews/5941578/Corbys-toxic-waste-the-families-speak-out.html. Accessed 13 Jan 2015.

Greenpeace. (2000). Thousands of radioactive waste barrels rusting: Greenpeace research unveils nuclear legacy. http://www.greenpeace.org.uk/media/press-releases/thousands-of-radioactive-waste-barrels-rusting-away-on-the-seabed. Accessed 6 Jan 2015.

Harvey, D. (2003). The new imperialism: On spatio-temporal fixes and accumulation by dispossession. *The Socialist Register, 40,* 63–87.

Higgins, P., Short, D., & South, N. (2013). Protecting the planet: A proposal for a law of Ecocide. *Crime, Law and Social Change, 59*(3), 251–266.

Howden, D. (27 April 2010). Visible from space, deadly on Earth: The gas flares of Nigeria. *The Independent, 2–3.* http://www.independent.co.uk/news/world/africa/visible-from-space-deadly-on-earth-the-gas-flares-of-nigeria-1955108.html. Accessed 19 Dec 2014.

Hunter, D., Salzman, J., & Zaelke, D. (2002). *International environmental law and policy, 2/e*. New York: Foundation Press.

Huo, X., Peng, L., Xu, X., Zheng, L., Qiu, B., Qi, Z., Zhang, B., Han, D., & Piao, Z. (2007). Elevated blood lead levels of children in Guiyu, an electronic waste recycling town in China. *Environmental Health Perspectives, 115*(7), 1113–1117.

Jessop, B. (2002). Liberalism, neoliberalism and urban governance: A state-theoretical perspective. *Antipode, 34*(3), 458–78.

Katz, R. (2010). The corporate crimes of Dow chemical and the failure to regulate environmental pollution. *Critical Criminology, 18*(4), 295–306.

Kostelnik, J. (2012). Sentencing white-collar criminals: When is shaming viable? *Global Crime, 13*(3), 141–212.

Liddick, D. (2010). The traffic in garbage and hazardous wastes: An overview. *Trends in Organised Crime, 13,* 134–146.

Lynch, M. (2013). Reflections on green criminology and its boundaries: Comparing environmental and criminal victimization and considering crime from an eco-city perspective. In N. South & A. Brisman (Eds.). *The Routledge international handbook of green criminology*. London: Routledge.

Mesothelioma Center. (2014). Asbestos mines. http://www.asbestos.com/occupations/mining.php. Accessed 13 Jan 2015.

McCarthy, J., & Prudham, S. (2004). Neoliberal nature and the nature of neoliberalism. *Geoforum, 35,* 275–283.

Morris, S. (2 Feb 2013). "Disaster" for UK seabirds hit by pollution. *The Guardian, 4*.

Neer, R. (2013). *Napalm: An American biography*. Cambridge: Harvard University Press.

Parmentier, R. (1999). Greenpeace and the dumping of waste at sea: A case of non-state actors' intervention in international affairs. *International Negotiation. A Journal of Theory and Practice, 4*(3), 435–457.

Pearce, F., & Tombs, S. (1998). *Toxic capitalism: Corporate crime and the chemical industry*. Aldershot: Dartmouth.

Pegg, S., & Zabbey, N. (2013). Oil and water: The Bodo spills and the destruction of traditional livelihood structures in the Niger delta. *Community Development Journal, 48*(3), 391–405.

Ringius, L. (2001). *Radioactive waste disposal at sea*. Cambridge: MIT Press.

Rose, D. (18 Nov 2007). "Safe" uranium that left a town contaminated. *The Observer,* 38–39.

Royal Society. (2012). The report on *people and the planet*, London: The Royal Society. https://royalsociety.org/policy/projects/people-planet/report/. Accessed 13 Jan 2015.

Ruggiero, V., & South, N. (2010). Green criminology and dirty-collar crime. *Critical Criminology, 18*(4), 251–262.

Ruggiero, V., & South, N. (2013a). Green Criminology and crimes of the economy: Theory, research and praxis. *Critical Criminology, 21,* 359–373.

Ruggiero, V., & South, N. (2013b). Toxic state–Corporate crimes, neo-liberalism and green criminology: The hazards and legacies of the oil, chemical and mineral industries. *International Journal for Crime, Justice and Social Democracy, 2*(2), 12–26.

Sellers, C., & Melling, J. (2011) *Dangerous trade: Histories of industrial hazard across a globalizing world*. Philadelphia: Temple University Press.

Shell Global. (2013). Flaring in Nigeria. http://web-dev2.shell.com/global/environment-society/society/nigeria/flaring.html. Accessed 19 Dec 2014.

Shell Companies in Nigeria. (2014). Gas flaring. http://s06.static-shell.com/content/dam/shell-new/local/country/nga/downloads/pdf/2014bnotes/gas-flaring.pdf. Accessed 13 Jan 2015.

Shrope, M. (28 Jan 2013). Minor oil spills are often bigger than reported: Remote imaging finds official number of Gulf of Mexico slicks is correct, but size not always is. *Nature: International Weekly Journal of Science*. http://www.nature.com/news/minor-oil-spills-are-often-bigger-than-reported-1.12307. Accessed 13 Jan 2015.

Simon, D. R. (2000). Corporate environmental crimes and social inequality. *American Behavioural Scientist, 43*, 633–645.

Smith, K., Fooks, G., Collin, J., Weishaar, H., Mandal, S., & Gilmore, A. (2010). Working the system–British American tobacco's influence on the European Union treaty and its implications for policy: An analysis of internal tobacco industry documents. *PLoS Medicine, 7*(1), e1000202. http://www.plosmedicine.org/article/fetchObject.action?uri=info%3Adoi%2F10.1371%2Fjournal.pmed.1000202&representation=PDF. Accessed 13 Jan 2015.

Sollund, R. (2008). Global harms: Ecological crime and speciesism. New York: Nova.

South, N. (2010). The ecocidal tendencies of late modernity: Trans-national crime, social exclusion, victims and rights. In R. White (Ed.), *Global environmental harm: Criminological perspectives* (pp. 228–247). Cullompton: Willan.

South, N. (2014). Green criminology: Reflections, connections, horizons. *International Journal for Crime, Justice and Social Democracy, 3*(2), 6–21.

South, N., & Brisman, A. (Eds.). (2013). *The Routledge international handbook of green criminology.* London: Routledge.

Tweedale, G. (2000). *Magic mineral to killer dust: Turner and Newall and the asbestos hazard.* Oxford: OUP.

United Nations Environment Programme. (8 Aug 2011). The environment in the news. (Monday, 8 August 2011). *United Nations Environment Programme*. www.unep.org/cpi/briefs/2011Aug08.doc; Accessed 13 Jan 2015.

Vidal, J. (30 May 2010). 'Nigeria's agony dwarfs the Gulf oil spill. The US and Europe ignore it. *The Observer*. http://www.theguardian.com/world/2010/may/30/oil-spills-nigeria-niger-delta-shell. Accessed 13 Jan 2015.

Vidal, J. (07 Aug 2011). Deep in the delta, the perfect village that died from a tide of oil pollution. *The Observer, 2*.

Vidal, J. (07 Jan 2015) Shell announces £ 55m payout for oil spills. *The Guardian*. http://www.theguardian.com/environment/2015/jan/07/shell-announces-55m-payout-for-nigeria-oil-spills. Accessed 13 Jan 2015.

Walters, R. (2007). Crime, regulation and waste in the United Kingdom. In P. Beirne & N. South (Eds.), *Issues in green criminology.* Cullompton: Willan.

Walters, R. (2010). Eco-crime. In J Muncie, D Talbot, & R Walters (Eds.), *Crime: Local and global* (pp. 173–208). Collumpton: Willan.

Walters, R., & Westerhuis, D. (2013). Green crime and environmental courts. *Crime, Law and Social Change, 59*(3), 279–290.

Week. (07 Sept 2012). The. 'Agent Orange's shameful legacy'. http://theweek.com/article/index/232816/agent-oranges-shameful-legacy. Accessed 13 Jan 2015.

White, R. (2008). *Crimes against nature: Environmental criminology and ecological justice.* Cullompton: Willan.

White, R. (Ed.). (2010). *Global environmental harm: Criminological perspectives*. Cullompton: Willan.

White, R. (2013). Environmental crime and problem-solving courts. *Crime, Law and Social Change, 59*(3), 267–278.

Zierler, D. (2001). *The invention of ecocide: Agent orange, Vietnam and the scientists who changed the way we think about the environment.* Athens: University of Georgia Press.

Dr. Nigel South is a Professor of Sociology at University of Essex in Colchester, UK.

Chapter 3
How E-Waste Challenges Environmental Governance

Lieselot Bisschop

Introduction

E-waste, waste from electronic and electrical equipment[1], is a fast growing market, with an estimated 72 billion tonnes generated annually worldwide by 2017 (Duan et al. 2013). Obviously, the digital (r)evolution has not only resulted in a significant increase in the quantity of e-waste but also a change in quality of the waste with the use of hazardous substances (Pellow 2007). E-waste is considered a hazardous waste because some of its components are harmful to the environment or to human health when not disposed off with care. Examples of such substances are heavy metals such as lead (e.g., in cathode ray tube, CRT screens and batteries), cadmium (e.g., in printer inks and toners), or compounds such as chlorofluorocarbon (CFC in cooling units).[2] When inappropriately disposed off, e-waste pollutes the soil, air, and water, thereby harming ecosystems, crops, and drinking water, causing harm to human health and the natural environment (Baker et al. 2004; Schluep et al. 2011). The precarious working circumstances in dismantling e-waste in West Africa and South East Asia have also been documented several times (Bisschop and Vande Walle 2013). Other types of harm relate to low material recovery rates, which is problematic in times when natural resources are scarce, and the unfair price settings, that disadvantage treatment companies that do adhere to environmental regulations.

The chapter originally appeared as an article in Vol. 3, No 2, pages 81–95 of the Special Edition on Green Criminology in the *International Journal of Crime, Justice and Social Democracy*.

[1] E-waste refers to all types of electrical and electronic equipment (EEE) that is discarded, such as television sets, computers, mobile phones, tablets, refrigerators, washing machines, dryers, home entertainment and stereo systems and toys.

[2] For more information, please consult http://ewasteguide.info/hazardous_substances.

L. Bisschop (✉)
John Jay College of Criminal Justice, New York, USA
e-mail: lieselot.bisschop@ugent.be

Ghent University, Ghent, Belgium

T. Wyatt (ed.), *Hazardous Waste and Pollution,* DOI 10.1007/978-3-319-18081-6_3

Waste is one of the most prominent environmental issues in contemporary society and at the same time one of the most normalized. The magnitude of the waste problem is significant and has increased with the pace of world population and consumption. As regulations on releasing waste in water, air, or land have become more stringent, the price of waste management services has increased (Vander Beken 2007). This has caused waste, including hazardous waste, to become an important article of trade. For instance, about 15 % of all trade within the European Union (EU) is trade in waste (IMPEL-TFS 2006). As for the cross-border or transnational dimension of the waste trade, most stays within the same region (e.g., EU) or takes place between countries that are members of the *Organisation for Economic Cooperation and Development* (OECD).

The reasons for the trade in (hazardous) waste are diverse. First, several countries do not have the necessary facilities to treat their hazardous waste or lack sufficient capacity to treat it all. Also, certain types of waste can only be dealt with in a limited number of facilities. Examples of these are the Swan Hills facility in Alberta Canada which treats dioxins and furan contaminated materials and the nuclear fuel recycling plant in La Hague France. Additionally, today West Africa lacks treatment facilities for hazardous waste. Second, sometimes a facility in another country is closer than one in the country of origin of the waste and hence transport across a border makes sense. Third, certain types of hazardous waste contain valuable secondary materials to be used in production processes in receiving countries. For e-waste, in particular, this issue of recycling and reusing secondary materials is prominent. Fourth, and perhaps most important, it is often less expensive to send waste to another country.[3]

Because of the abovementioned harms resulting from inadequate recycling or disposal of waste, the trade in waste is regulated in multilateral environmental agreements.[4] The trade flows that are most likely to result in inadequate recycling or disposal are those from the global North (e.g., EU, the USA, and Australia) to South (Africa, South East Asia, and South America) (Basel Action Network and Silicon Valley Toxics Coalition 2002; Basel Action Network 2005; Crem 2008). In fact, the illegal trade in and disposal of hazardous waste has been identified as a major form of environmental crime. For instance, under European law it is illegal to export toxic waste to non-EU countries with weaker standards for disposal. Although exact data do not exist, results of enforcement actions coordinated by the *European Union Network for the Implementation and Enforcement of Environmental Law* (IMPEL) give some indication of the scale of the violations (IMPEL-TFS 2012). Out of the

---

[3] http://www.cec.org/hazwaste/Page.asp?PageID=5001&ContentID=341&SiteNodeID=204&BL_ExpandID=60.

[4] The discovery of and media attention for dump sites of toxic waste in developing countries during the 1980s and 1990s led to the adoption of international and European legislative frameworks that regulate waste transports. Examples of these are the Basel Convention on the control of transboundary movements of hazardous wastes and their disposal; Montreal Protocol on ozone-depleting substances; and OECD decision on control of cross-border movements of waste destined for recovery operations; European Waste Shipment Regulation; Directive on Waste Electrical and Electronic Equipment; Bamako Convention on the Ban of the Import into Africa and the Control of Transboundary Movement and Management of Hazardous Wastes within Africa.

total number of waste inspections carried out, 28.9 % were either an administrative violation, which refers to missing or incomplete forms (e.g., Annex VII), or an illegal shipment which refers to a transport that is subject to an export ban.[5] Waste from electronic and electrical equipment (WEEE) is one of the major waste streams among these transport violations. Overall, one in five containers exported from the EU contain waste and an estimated 20 % of those are in violation of export bans or administrative requirements for waste trading (Baird et al. 2014). However, the data available on illegal trade in e-waste reflect the control that takes place and are not necessarily an accurate reflection of the scale of smuggling (Bisschop 2012).

Against this background, this chapter aims to shed light on how e-waste challenges environmental governance. Governance differs from government because it is not limited to the engagement of state actors (Holley et al. 2012; Parker and Braithwaite 2003). Governance refers to different social and political units that govern social interactions such as corporations, international organizations and nongovernmental organizations (NGOs). Governance, in this chapter, equals those initiatives that aim to prevent the harm of substandard e-waste disposal and to control illegal e-waste trade. Throughout this analysis, this chapter relates back to theories about environmental governance and responsive regulation (Holley et al. 2012; Parker and Braithwaite 2003).

The following section discusses the methodology of the study described and sketches the characteristics of the research context. The chapter continues by discussing the challenges inherent to e-waste both as a commodity and as a sector and then explains the challenges accompanying legal definitions and policy-making. The focus then shifts to the challenges inherent in the implementation of policy regarding e-waste and the difficulties that emerge in various interactions between governance actors in preventing harm and controlling crime related to e-waste. Attention is paid to the importance and difficulty of involving local governance actors, corporations, and nongovernmental organizations in governing e-waste. This chapter does not provide an exhaustive list of the governance challenges in dealing with e-waste but focuses on a number of core themes. This, nevertheless, demonstrates the rich complexity of the reality of governance.

Method and Research Context

This chapter draws on the findings of a case study describing the trade of e-waste passing through the port of Antwerp in Belgium (Bisschop 2013, 2015). The flows between Belgium and Ghana were the particular focus. The analysis is thus tailored to that specific empirical case. Antwerp, with a total yearly freight volume of 185 million tons is among the top three ports in Europe alongside Rotterdam and Hamburg. Antwerp is a landlord port, owned by government and managed by a port

[5] Over the several phases of IMPEL-TFS measurements (2006–2012), the number of administrative violations decreased from 52 to 37 %, but the number of illegal shipments, subject to an export ban increased from 33 to 38 %.

authority which gives up its control over port operations by leasing the infrastructure to private terminal operators (Talley 2009). The Antwerp port is a complex area for prevention and control covering 130 km^2, 180 km of quay, 400 km of roads, 1000 km of railway, and with about 900 companies working in the port area. The territory of the port of Antwerp is governed by two jurisdictions (Dendermonde and Antwerp) and three municipalities (Antwerp, Beveren, and Zwijndrecht) and falls under the responsibility of two provincial governments (East-Flanders and Antwerp). Given the sheer volume of trade, Antwerp employs a risk analysis system to control transport. Less than 1 % of containers are scanned. About 80 % of the waste that is traded in the port is in transit, the other 20 % has its origin or destination within the territory of Belgium.

The port of Tema handles over half of Ghana's trade and is also an important gateway for Burkina Faso, Mali, and Niger. In 2012, it handled 11.5 million tons of freight, a 50 % increase compared to 2008. Imports account for about 80 %, exports 15 % and transit 5 % (Eshun and Rasmussen 2013). All containers arriving in Tema are either scanned or physically inspected by unloading them.

This chapter makes no claim about being able to generalize from these findings to the cases of other regions and trade flows. By analyzing the case within its broader context and paying attention to the global trade flows, it does provide insights that help to understand the challenges to the governance of illegal trade in and substandard dismantling of (e-)waste in other locations.

The findings are based on a multi-method approach of document analyses, interviews, and field visits. Both the document analyses and the interviews triangulated government, corporate, and civil society perspectives. The document analyses studied the existing international, European, and national (Belgian and Ghanaian) regulation and litigation as well as available corporate and NGO documents. This study was further based on 56 semi-structured interviews with law enforcement (customs, police, environmental inspectorates, port authority, and judiciary), corporate actors (shipping lines and agents, e-waste collectors and recyclers), and civil society actors (informal e-waste workers, NGOs, and local communities). The findings from the document analyses and the interviews were triangulated with data from the field visits in the port of Antwerp (Belgium), the city of Accra, and the port of Tema (Ghana).

Challenged by Waste as a Product and a Sector

There are a number of problems that relate to waste as a product. It has been linked to price fixing, racketeering, and an illegal market in waste (Van Daele et al. 2007). It has an inverse incentive structure due to its negative value: when you own waste you usually need to pay to get rid of it. This inverse incentive structure does not straightforwardly apply to e-waste because there may often be some value in owning it as the (precious) metals it contains are worth retrieving for resale. Waste is also generally considered as a product of low integrity since it can be easily mixed up or sold as second hand commodity (Gibbs et al. 2010; Van Daele and Vander Beken 2009). The transition from being "legal" to "illegal" when dealing with waste can occur at several stages. This can happen not only during national and cross-border transportation but also in collection or disposal. Waste can be more cheaply dealt with by illegal enterprises who disregard environmental regulations or by legal companies in Europe who treat waste they are not licensed for (Ruggiero and South 2010).

As well as the product itself attracting illegal activity, the waste collecting and processing sector as a whole has characteristics that can be considered vulnerable to crime (Van Daele et al. 2007). Its rapid growth and international character have allowed a diverse range of actors to work in the waste industry, with brokers involved at different stages obscuring the paper trail. In collection, transport, and treatment, multiple smaller companies try to compete with few big ones. Many substreams of waste arise from the dismantling of various products and multiple actors are involved. Corporations that process e-waste legally and have environmentally sound management systems are few and their prices are high. The category of e-waste includes many devices that incur costs for treatment before disposal but some products hold enough valuable components to make recycling or treatment profitable (LNE and Haskoning 2010). For the former, exporting them as secondhand products saves the cost of treatment (Interpol 2009). The latter is a motivation for having the goods dismantled as cheaply as possible to be able to sell the raw materials (Sander and Schilling 2010). "You can make money by 'recycling' e-waste in poorer environmental and social conditions, since this provides you the precious metals with lower labour costs" (Corporate respondent 13).

The concern for the loss of secondary raw materials through e-waste exports is very important for the EU as it relies on the inflow of e-waste for the operation of its high-standard recycling facilities. Some recyclers see the setting of stringent legislation on waste exports as profitable for their business. Some refurbishers set even more stringent standards for their exports of used electronic and electrical equipment (UEEE) than the law requires. Evaluations and stakeholder consultations are continuing with the aim of developing a certification scheme or standard to demonstrate that waste exported from the EU to non-OECD countries can be treated in an environmentally sound manner. This usually involves firms that have made this their niche market. As long as all parties can benefit in a "win-win" situation, investments seem to be made but there are also practical limitations (Gunningham et al. 2003). For example, producers were found to be cherry picking in terms of their responsibility to take back WEEE from Ghana. Indeed, willingness and ability to commit to self-regulation and monitoring are investments largely limited to the "big players" or niche corporations (Gunningham et al. 2003). This, of course, leaves a multitude of smaller corporations where self-regulation is much less clear (Haines 1997). Nonetheless, in the supply chain of electronics and e-waste and the due diligence that lies within, smaller actors can be motivated to set high standards for themselves. Some large corporations in collection and recycling choose which influx of WEEE/UEEE to accept, depending on whether they deem the source trustworthy. There is thus room to take this a step further and to truly integrate the environmental costs throughout the entire waste processing life cycle (van Erp and Huisman 2010). To address this, extended producer responsibility policies have emerged[6] (Pellow 2007) but the implementation of these can still be improved. Both producers of electronics and the waste sector can take steps to prevent environmental harm in the supply chain of electronics and e-waste. Although both the product and the sector represent a challenge, corporations have the potential to play a role in environmental governance. Major corporations, such as producers of electronics and (e-)waste processors, have taken the initiative to self-regulate, albeit with varying degrees of success.

The governance of the illegal trade in e-waste, to a certain extent, is addressed through a form of "regulated self-regulation". This, for instance, happens in the risk analysis system employed by European customs which bases its controls on the management systems of the corporations besides other risk indicators[7] informed by legislation and past experience of customs, environmental inspectorates, police, and administrations. Similarly, the inspectorates work through meta-regulation to control waste corporations. Self-regulation has the advantage of being more attuned to the underlying processes of the waste sector rather than having only the shallow effect that may follow when only the outcomes are controlled (de Bree 2011). Self-regulation might also be more flexible in addressing the dynamic reality of preventing illegal transport of e-waste (e.g., perceiving CRT television sets as WEEE). Self-regulation cannot address the entire market of waste collection and processing, given the multitude of small scale actors (Bisschop 2012) which make it difficult to install self-regulation sector-wide, for instance, through certification schemes or extended responsibility (Gunningham et al. 1998; Holley et al. 2012). Setting this up through (regulated) self-regulation implies that all parties need to agree on the purpose and benefits (White 2011).

Challenged by Legal Definitions and Policy-making

Many international organizations have developed policy mechanisms to manage waste generation and regulate the trade in waste (Iwama 2004).[8] As it is subject to international conventions, the trade in hazardous waste can be conceptualized in terms of legality and illegality. The most important conventions are the 1989 Basel Convention on the Control of Transboundary Movements of Hazardous Wastes and their Disposal and its 1995 ban amendment on trade of Waste for recovery between

[6] For instance the European WEEE Directive (on waste of electronic and electrical equipment) or the Directive on the restriction of hazardous substances.

[7] These risk indicators can refer to particular destinations (risk countries, suspicious streets), descriptions of the goods, value added tax (VAT) numbers and the value of the goods.

[8] For instance: UNDP (solid waste management); WTO (sustainable/recycling waste trade); World Bank (solid waste management); OECD (waste management, sustainable use of materials); (2001); WCO (combating environmental crime—MEAs); INTERPOL (illegal waste trade).

OECD and non-OECD countries, the 1987 Montreal Protocol on Substances that deplete the Ozone layer and the 2001 OECD decision concerning the Control of Transboundary Movement of Wastes destined for Recovery Operations. Basically, the more hazardous a waste shipment is, the more legal requirements it needs to answer to such as providing information about the facility it is destined for or prior consent of receiving country before the transport takes place. Waste crimes are considered an important type of environmental crime and make up a large share of the cases reported by environmental inspectorates. These can refer to both administrative (missing or incomplete paperwork) and criminal (going against actual trade ban) violations.

The policy framework itself is vulnerable to criminality in several ways. The Basel Convention and EU legal framework for waste are not always interpreted and implemented with the same commitment among all the parties. Not all countries have ratified the Basel Convention and its Ban Amendment. Moreover, the Basel Convention is primarily intended to regulate trade and not to control illegal trade. National differences in policy and particularly, in implementation still exist, even within the EU. A particular difficulty is distinguishing the *illegal* exports of e-waste to non-OECD countries and the *legal* exports of secondhand electronics. The latter need to meet the following requirements: proof of functionality, contact details of shippers and destination, and appropriate packaging. Despite these guidelines, it can still be challenging to distinguish the two. Overall, the EU waste legislation is very complex. As a consequence of this, customs, harbormaster office, and police rely on environmental inspectorates to judge the suspicious shipments. Even prosecutors admit the legal framework for waste is difficult to apply. This confirms the finding that waste is a topic with many rules and regulations that are not always practical (Huisman 2001). This goes against one of the basic principles of good governance that it needs to be understandable (Braithwaite 2008).

What complicates this further is the dynamic nature of what can be defined as e-waste. There is a temporal dimension in the sense that more knowledge about harmful substances emerges daily, but another dynamic is also at play in the sense that the discussion about what is a secondhand electronic product and what is e-waste depends on the cultural and socioeconomic context. Belgian definitions of e-waste are different from Ghanaian and therefore definitions of environmental crime and harm also differ. Similar to other issues of environmental harm, the definition of what constitutes illegality varies over time, place, and perspective (White 2011). It is not easy for policy and legislation to incorporate this dynamism.

Challenges in Implementation

Despite e-waste fraud being an issue of national and international importance, resources are limited as is political will (Griffiths and Jenks 2012). Both in countries of origin and destination, there are a limited number of people involved in detection, despite the multitude of government agencies dealing with it. Environmental inspectorates play a crucial role and usually have the most technical expertise about waste matters. Environmental inspectorates are also those with most knowledge about waste collection, treatment, or disposal facilities. This means that they can address companies on their license in cases of administrative violations and might be able to trace illegal waste trade back to its source. There is, however, a serious lack of capacity for administrative follow-up of illegal waste transports, since inspectorates have limited staff. Police officers focus on environmental issues and lead criminal investigations into waste fraud but, in much the same way, they are limited in staff and resources. One way to come across any illegal transport of e-waste is through the control of shipments in harbors. To a certain degree, this is like looking for a needle in a haystack. Customs play an important role in this through their risk analysis system for scanning and physically inspecting shipments. However, environmental issues are not a priority for customs, resulting in limited support from the hierarchy. Furthermore, the problem requires technical expertise that is not provided for in training. Despite their knowledge and experience being crucial for the control of waste transports, enforcement agencies are bound by their limited resources. "The problem is not that we don't know where the illicit traffic is, know where to find it or know how to check it. The problem is that we don't have the means to guarantee the follow-up, that's the bottleneck"(Government respondent 14). This refers to those environmental inspectors and police officers who are governing e-waste transports on a daily basis. These exceptions to the rule are passionate about their job but frustration does occur.

An additional challenge of EU waste policy is the different implementation in the EU member states. The case of the port of Antwerp illustrates this perfectly. Belgian inspectorates who control shipments in the port, can only trace illegal waste shipments to their source when they originated in Belgium. As mentioned earlier, this is about one fifth of all waste transports in the port. For the other 80 % of waste—waste in transit from other EU countries—the exporting member states are responsible for ensuring that the goods accord with legislation. Similarly, the risk analysis system of Antwerp customs is applied to import and export, but the transit shipments rely on the risk analysis of the other EU countries. However, these member states do not necessarily interpret or implement the waste legislation in the same way. Not all member states use complementary interventions as well as the traditional environmental inspection, and when controls become more severe in one country, illegal flows shift to another. There is no level playing field for controls on waste in the EU. Government, corporate, and civil society respondents even notice a North–South and West–East schism in Europe concerning the matter of how seriously inspections and transport controls are taken. This led different respondents to refer to the need for further harmonization of EU policy implementation to avoid displacement, although this is politically very sensitive. Imposing minimum requirements for inspections and controls is often interpreted as too much interference from Europe, especially if that would imply countries can call each other to order. Recently, the European Parliament confirmed an agreement to revise the EU rules on waste shipments. "The new law will ensure that member states are obliged to make comprehensive and meaningful inspection plans to check waste shipments, with a minimum number of physical checks in line with the risk of illegal shipments. … Enforcement will be stepped up, with more powers for authorities involved in inspections to check shipments. Cooperation between member states will be significantly strengthened." (MEP Bart Staes, 17 April 2014)[9]. Taking this a step further, setting up a European environment agency with inspectorate authority might be beneficial to organize the controls at the EU level and create a more unified system, but this is a politically very sensitive topic.

Even if definitions and interpretations would be equal, Europe—let alone global policy—is a long way from harmonization of implementation. Prosecution of environmental crime remains a national competence with significant differences between European countries in approach and number of convictions (IMPEL-TFS 2013). To avoid criminals from taking advantage of this, IMPEL identified the establishment of a network of prosecutors for exchange of relevant case law, prosecution information like the level of fines, working methods, prosecution approach, interpretation, and practical experiences. Although the European waste legislation requires sanctions to be set, there are major differences in interpretation and implementation. There is prosecution of waste cases but fines that are imposed for illegal e-waste transports are perceived as too low to be effective and become part of shippers' business plans. One of the reasons for this is that, although it is often possible to prove one shipment is illegal, it may be hard to prove this has happened systemically. Fines differ with a factor of 100 across the EU. Even within countries there are considerable differences in decision-making (Sander and Schilling 2010). This is similar to findings about the implementation of other environmental policies in the EU, where member states have a great margin of interpretation in their determination of sanctions (Billiet and Meeus 2010). To counter this, judges could be guided about the seriousness of transnational environmental crime (White 2011) as suggested by respondents in this study.

Also on the receiving end of the trade flows, in countries of destination like Ghana, the implementation of legislation can prove challenging. Ghana has signed international conventions but national legislation on e-waste is not yet in place, despite a proposal for a bill on the control and management of hazardous waste in general, and on e-waste in particular, being drafted in 2012. In the absence of a legal basis to act, the main concern of the Ghanaian Environmental Protection Agency (EPA) is raising awareness of government actors about the dangers of e-waste since many consider them as profitable secondhand products. The enforcement focus is on licensing recycling facilities and analyzing where WEEE/UEEE might be sold and refurbished. However, these actors are numerous, very flexible and spread out across Ghana. The EPA also reaches out to informal actors and shows them how to improve working conditions. In theory, customs have governance potential in tackling the illegal e-waste transports since they inspect and unload containers of WEEE/UEEE. However, their priority is taxing the import of electronics rather than checking for e-waste. In case a law is passed, awareness raising and training of these enforcers will be crucial. Even then, the problem remains that seized goods will have to be recycled locally, often ending up in the same system. Increased attention to the trade in e-waste also does not provide a solution to processing the large amounts of new and secondhand electronics being consumed in Ghana.

[9] http://www.greens-efa.eu/waste-shipments−12312.html [Last consulted 24 April 2014].

Challenges in Involving Local Governance Actors

Overall, the governance framework seems to be government dominated for e-waste, whereas corporate and civil society actors are more passive. When the authority over a port area is spread across different terminal operators, municipalities, provinces, and jurisdictions, this renders governance complex, but in recent years, improvements have been made to step up the cooperation and information exchange among these actors in the port of Antwerp. In examining the governance reality of e-waste, the primary actors in the control and prevention of illegal transport remain the inspectorates. In theory, these environmental inspectors can trace transports in violation of either administrative or criminal regulation to its source, usually being a particular corporation involved in waste transport, collection, treatment, or disposal. As mentioned earlier, tracing the transit shipments to their sources remains the responsibility of governments in other member states. This approach of tracing the waste to its source does not easily apply to the multitude of small-scale (often individual) shippers. Small-scale collectors are not always known to the environmental inspectorates. To be more exact, it might be a shipment by one individual and not by a registered waste corporation that went through the licensing process for its facility. In preventing illegal waste shipments, it is however important to also get a grip on these smaller scale actors. Relying too much on the regulatory activities of inspectorates, transport controls by environmental police officers, and the controls in harbors by customs or harbormaster's office, risks not getting a grip on the more local level of waste smuggling. In identifying those smaller scale transports, it is important to involve local governments and local police organizations because they might have a better view on waste collection activities—of whatever scale—that happen in their communities. This would allow tackling the problem by starting at the root, focusing on actors earlier in the supply chain of electronics and waste.

The local level of governance is, therefore, important for the success of governing environmental flows despite their transnational characteristics (Braithwaite 2008; Gille 2006). This is particularly relevant when the primary concern is shifted from controlling of crime to avoiding environmental harm. In developing countries, promising local initiatives are those of collecting and dismantling of e-waste (e.g., in Ghana), where scrap dealers are taught techniques to dismantle equipment while reducing health and environmental risks, ultimately resulting in better recovery rates for the raw materials. These initiatives are still very few, usually setup with the support of NGOs. Local Ghanaian NGOs also combine forces with international organizations to develop capacity building. Similarly, they cooperate with (European) corporations for the return of equipment (e.g., motherboards). Some companies are also looking to follow this example for other components, but are assessing the economic feasibility. There are initiatives to turn the highly efficient informal economy of waste collection into a less harmful waste industry that provides extra income for the population and prevents further pollution (prevention of hazardous waste leaking into the soil or prevention of burning hazardous waste). Although many of these initiatives are still in their infancy, this shows that a broader network exists, beyond law enforcement alone, for dealing with e-waste. Setting up these initiatives takes time because local characteristics need to be taken into account. Insights are needed about how and why these governance initiatives work. Only then is it possible to evaluate whether it could apply elsewhere. This also implies the need to avoid imposing a Western frame of reference without considering local differences.

Challenges in Interactions Between Governance Actors

Given the above challenges presented by legal definitions and implementation of waste legislation, it seems fitting not to rely solely on the criminal justice system to prevent environmental harm from inadequate disposal of e-waste and control illegal transports of e-waste. Complex problems such as environmental issues are not easily governed by a single actor and a networked governance model might fare better (Holley et al. 2012). In interaction, however, it is not always easy to work towards the same goals, even amongst government agencies—as the e-waste case demonstrates. The governance of e-waste cuts through different layers of authority. Each of the government actors who were examined in this study were aware that their work was only one part of a broader system. Depending on the features of the case, one or another actor took a leading role, either for practical reasons, due to available expertise or out of concern for an integrated approach. The judiciary clearly takes a leading role in the approach to waste fraud in the port of Antwerp. Customs take a leading role in the risk analysis system. Inspectorates and administrations have expertise on environmental issues. Although these actors are all part of the network or the chain that responds to these issues, their methods and responsibilities are different. One of the criticisms of corporations is that the implementation of the law seems to be valued higher than the matter of environmental importance. Corporations experience an extra burden in terms of administration because of the fragmentation of agencies and perceive this as an inefficient use by government of already limited environmental enforcement resources. This efficiency of the network will become increasingly important, given the continued expansion of the Antwerp harbor and its trade flow. This fragmentation of responsibility is also being addressed in Belgian institutional reforms.

Waste is a topic, just like other environmental issues, that can be dealt with by administrative as well as criminal proceedings and both do not necessarily have the same objectives in mind. Transparency and clear delineation of tasks and objectives is crucial to avoid government actors fighting each other instead of fighting crime (Sluis et al. 2012). The question then is, whether these actors work with the same intentions in mind and the same realities at hand. Environmental inspectorates focus on illegal transports of e-waste but also take into account the broader concern of avoiding environmental harm. Administration systems are similarly oriented to environmental issues but focus on initiatives earlier in the chain (licenses) as well as on policy. The focus of customs and harbormaster's office on illegal trade in e-waste is rather recent and struggles with balancing concerns for security and the economy. Customs specializes in responsibilities that are at the front line in import and export of waste; it has the technical means to control (scanning of containers and risk analysis) and has many “eyes and ears”. Note, however, that their risk analysis system which relies on a large basis of trust in corporations (potentially) does not match the approaches of other government actors. Questions remain whether this system of trust in economic operators by customs and even the system of trust in compliance by the environmental administrations can be united with the strategies of judicial actors that proceed in their work based on distrust. The police and the judiciary are both concerned with illegal transport of e-waste because they are environmental crimes and have a potential link to “organized crime”. This only concerns the serious and extensive cases.

Besides the various initiatives for international networking, the responsible authorities cooperate bilaterally with their neighboring countries as well as with countries of destination such as Ghana, but with limited staff and resources this can be difficult. Different international guidelines and soft measures exist and are stimulated by organizations such as the IMPEL, International Network for Environmental Compliance and Enforcement (INECE), Secretariat of the Basel Convention, the World Customs Organization (WCO), and the Solving the E-waste Problem (StEP) initiative. These initiatives organize training sessions in countries of destination, function as advisors, and fund capacity-building projects. One challenge to these international networking initiatives is their voluntary character. Unambiguous measures (e.g., inspections and controls) are more difficult to agree on, though, this should not be surprising given the known problems of global environmental governance (Iwama 2004): an ad hoc or fragmented approach, a missing link between policy making and implementation, conflicting or overlapping responsibilities between different agencies, a lack of a powerful competent authority, and NGOs and corporate actors not being reflected in multilateral environmental initiatives. Despite significant improvements in policy making and—more recently—implementation, many of the above problems apply to the governance of e-waste.

From a governance perspective, nongovernment actors have a valuable role to play, although this is currently not exploited to its full potential in the case of e-waste. Both corporations and civil society stakeholders come to mind. When taking the supply chain of electronics into account, producers, recyclers, transport actors, and consumers become part of the governance framework. Not all producers, recyclers, and transport actors are currently proactive in self-regulating their sector, despite the governance potential they may have. For instance, transport actors could be encouraged to be more diligent and transparent. The first important step is raising awareness about this. Many company brochures and websites already mention sustainability and related issues, but there is a need for further translation of awareness into practice in both Belgium and Ghana. As witnessed during the field visits and interviews, many port actors do not know what e-waste is and are unaware of the potential harms. “This information campaign should contain practical advice directed not only at the managers of these companies but also at the actual goods handlers. This might not stop crafty criminals from shipping e-waste illegally but might address ‘low hanging fruit’” (Civil society respondent 20). This will likely require a few major shipping lines to take the lead, as is happening today, but it might be more difficult to involve smaller actors like shipping agents, "because these often have neither the resources nor the management structure to seriously deal with this" (Corporate respondent 5). A second step might be to increase information exchange about the flows—both legal and illegal—providing better grounds on which to address illegal trade. As a testimony to the governance potential of shipping lines, they are already partners of the waste enforcement agencies in England and Wales, thus helping to keep their business clean and trace waste back to its origin. It is, however, not possible to simply replicate this in Antwerp, where the share of transit in the waste trade is much larger (80 % in Antwerp versus 10 % in England and Wales). A similar development is occurring in countries of destination where shipping lines are increasingly providing the authorities with information to facilitate controls or even providing warnings in cases where suspicions about the nature of the shipment arise.

By phasing out hazardous components, producers can ensure that the recycling of e-waste is less harmful. Through eco-design, they can increase the life cycles of products. However, economic and environmental interests do not always coincide. Consequently, the involvement of producers and recyclers is largely determined by the potential gains from being environmentally responsible, either financially, or in terms of their corporate image. An important governance potential, however, lies in linking the illegal transports of e-waste to the EU's lack of raw materials. This is an important incentive for European industry, especially since it is often more energy-efficient to extract precious metals from e-waste than from mines. Protecting the secondary raw materials and keeping these within Europe is increasingly going hand in hand with the objective of avoiding further environmental degradation in countries of destination of e-waste transports. Despite these self-regulatory efforts, corporate actors could play a more significant role, for instance, through regulated self-regulation in the governance of illegal e-waste trade. This will, of course, require carefully designed incentives—both positive and negative (Gunningham et al. 1998; Holley et al. 2012).

Apart from the role for corporations, the civil society representatives like NGOs play a role in governance. This happens by raising consumer awareness and by keeping governments and corporations on their toes about environmental topics. During the field visits, it also became clear that NGOs play an important role in engaging local actors in Ghana. By setting up capacity-building projects, they are reaching out to actors who are currently not always involved in governance. Many of these initiatives are small-scale and ad hoc, but have the potential to address the structural causes of the illegal trade in e-waste. They can, for instance, aim to reduce the demand for WEEE/UEEE by highlighting the fact that dismantling e-waste and secondhand electronics in search of precious metals is the sole source of a secure livelihood for many people in countries of destination. More importantly, they can avoid the immediate harm by teaching informal workers about more environmentally and health-friendly dismantling techniques. This would address both, illegal trade in e-waste and the domestic consumption of electronics.

Conclusion

This chapter set out to shed light on how e-waste is a challenge for environmental governance. Challenges inherent in legal definitions and policy-making remain, outnumbered only by the challenges inherent in the implementation of waste policy. The analysis of the reality of governance illustrates how the control of illegal e-waste flows is primarily taken up by government actors, which is not surprising given the criminalization of it. An important challenge, however, is the underfunding of these government actors, with consequences for training, resources, and effective follow-up throughout the flows (See also Hayman and Brack 2002). Despite the good intentions of many and the improvements made in the past decade, both the administrative and the criminal approach in dealing with illegal transports of e-waste could be improved. Even in a region like the EU that can be considered an environmental forerunner (Vig and Faure 2004), environmental law enforcement is perceived to be too slow and too weak in responding to environmental crime. An important characteristic of responsive regulation is the escalation towards more punitive measures in cases of (continued) noncompliance (Braithwaite 2008; Nielsen and Parker 2009). As has been illustrated here, this escalation is not at all certain for the illegal trade in e-waste, not in the countries of origin and even less so in the countries of destination. For environmental governance to work, it is crucial to gain insights into the objectives and realities of each separate governance actor, to understand basic assumptions that might be influencing their cooperation. This might require trying to overcome differences and working towards the same end, despite different objectives and means. This also requires clarity about whether avoiding environmental harm or preventing environmental crime is the primary concern. Through a networked governance approach different interests can be taken into account and negotiation can proceed based on each actor's capacity. This means taking into account not only what is strictly defined as illegal but also the broader dimensions, allowing a more complete account of social phenomena. The diversity of actors and reasons to get involved in illegal e-waste transport or substandard treatment of e-waste, inevitably means taking many actors into account in countries of origin, transit, and destination and on local, national, and international levels. This, of course, renders the arrangement of networked governance very complex, going against the idea that networked governance for the environment is easiest in a small-scale setting with limited complexity (Holley et al. 2012).

A more comprehensive governance picture emerges when the gaze is broadened from controlling of illegal transports of e-waste to preventing environmental harm through substandard recycling practices. This makes it possible to look for governance frameworks that involve actors throughout the supply chain of electronics and not merely those that contribute to or control for (illegal) trade in e-waste. The emergence of new actors, outside of the state, on local levels and in new cooperative arrangements fits a view on "world politics that are no longer confined to nation states but are characterized by increasing participation of actors that have so far been largely active at the subnational level" (Biermann and Pattberg 2008, p. 280).

Insight is needed regarding the potential and challenges of these new environmental governance arrangements. In the interaction between government, corporate and civil society on the topic of e-waste, a challenge remains in the need to be clear about objectives and work towards similar ends despite differences. Also, the different perspectives of the global North and South will require environmental governance mechanisms to assess challenges and strengths within particular local contexts. At the same time, the inherently global dynamics of the trade in electronics as well as of environment harm require governance mechanisms that cut across national and institutional boundaries. A locally grounded and globally informed governance network holds the potential to prevent further environmental harm caused by the supply chain of electronics.

References

Baird, J., Curry, R., & Cruz, P. (2014). An overview of waste crime, its characteristics, and the vulnerability of the EU waste sector. *Waste Management & Research, 32*(2), 97–105.

Baker, E., Bournay, E., Harayama, A., & Rekacewicz, Ph. (2004). Vital waste graphics. United Nations Environment Programme, Basel Convention, GRID Arendal, DEWA Europe.

Basel Action Network. (2005). *The digital dump. Exporting re-use and abuse to Africa*. Seattle: The Basel Action Network (BAN).

Basel Action Network, & Silicon Valley Toxics Coalition. (2002). *Exporting harm: The high-tech trashing of Asia*. Seattle–San Jose: The Basel Action Network (BAN)–Silicon Valley Toxics Coalition (SVTC).

Biermann, F., & Pattberg, P. (2008). Global environmental governance: Taking stock, moving forward. *Annual Review of Environment and Resources, 33*(1), 277–294. http://www.annualreviews.org/doi/abs/10.1146/annurev.environ.33.050707.085733. Accessed 22 March 2014.

Billiet, C., & Meeus, R. (2010). Europese verordeningen en de handhaver. De sanctieregelingen van milieuverordeningen in het licht van de handhavingspraktijk. *Tijdschrift voor Milieurecht, 19*(2), 164–202.

Bisschop, L. (2012). Is it all going to waste? Illegal transports of e-waste in a European trade hub. *Crime, Law and Social Change,* 1–29. doi: 10.1007/s10611-012-9383-0.

Bisschop, L. (2013). Go with the e-waste flows. The governance reality of illegal transports of e-waste in a European trade hub. In P. van Duyne, J. Harvey, G. Antonopoulos, K. von Lampe, A. Maljević, & J. Spencer (Eds.), *Human dimensions in organised crime, money laundering and corruption* (pp. 393–424). Nijmegen: Wolf Legal Publishers.

Bisschop, L. (2015). Governance of the illegal trade in e-waste and tropical timber: Case studies on transnational environmental crime. Green Criminology Series: Ashgate Publishing.

Bisschop, L., & Vande Walle, G. (2013). Environmental victimization and conflict resolution. A case study of e-waste. In R. Walters, D. Westerhuis, & T. Wyatt (Eds.), *Emerging issues in green criminology: Power, justice and environmental harm* (pp. 34–54). Basingstoke: Palgrave Macmillan.

Braithwaite, J. (2008). *Regulatory capitalism: How it works, ideas for making it work better*. Cheltenham: Edwar Elgar.

CREM. (2008). *Een analyse van stromen electronica-afval in Nederland [Analysis of e-waste flows in the Netherlands]*. Amsterdam: CREM Milieu Onderzoeks- en Adviesbureau.

De Bree, M. (2011). Ontwikkelingen in systeemtoezicht. In M. de Bree (Ed.), *Managementsystemen en toezicht* (pp. 51–60). Erasmus Universiteit Rotterdam: Erasmus Instituut Toezicht & Compliance.

Duan, H., Miller, R., Gregory, J., & Randolph, K. (2013). *Quantitative characterization of domestic and transboundary flows of used electronics. Analysis of generation, collection, and export in the United States, MIT, MSL, NCER*. http://www.step-initiative.org/files/step/_documents/MIT-NCER%20US%20Used%20Electronics%20Flows%20Report%20-%20December%202013.pdf. Accessed 29 May 2015.
Eshun, A., & Rasmussen, N. (2013). *Trends in transport and logistics on the tema-ouagadougou-bamako corridor*, United State Agency for International Development, West Africa Trade Hub. http://www.watradehub.com/sites/default/files/Trends%20in%20Transport%20May%202013.pdf. Accessed 12 June 2014.
Gibbs, C., McGarrell, E. F., & Axelrod, M. (2010). Transnational white-collar crime and risk. Lessons from the global trade in electronic waste. *Criminology & Public Policy, 9*(3), 543–560.
Gille, Z. (2006) Detached flows or grounded place-making projects? In G. Spaargaren, A. Mol, & F. Buttel (Eds.), *Governing environmental flows. Global challenges to social theory* (pp. 137–156). London: MIT Press.
Griffiths, H., & Jenks, M. (2012). *Maritime transport and destabilizing commodity flows*. Solna: Stockholm International Peace Research Institute.
Gunningham, N., Grabosky, P., & Sinclair, D. (1998). *Smart regulation: Designing environmental policy*. Oxford: Clarendon Press.
Gunningham, N., Kagan, R., & Thornton, D. (2003). *Shades of green: Business, regulation and environment*. Stanford: Stanford University Press.
Haines, F. (1997). *Corporate regulation: Beyond "punish or persuade."* Oxford: Clarendon Press.
Hayman, G., & Brack, D. (2002). International environmental crime: The nature and control of environmental black markets. *International Affairs*, (May), pp. 27–28.
Holley, C., Gunningham, N., & Shearing, C. (2012). *The new environmental governance*. Abingdon: Routledge.
Huisman, W. (2001). *Tussen winst en moraal: Achtergronden van regelnaleving en regelovertreding door ondernemingen*. Den Haag: Boom Juridische Uitgevers.
IMPEL-TFS. (2006). IMPEL-TFS Seaport Project II: International cooperation in enforcement hitting illegal waste shipments. Project report Septemer 2004–May 2006, European Union Network for the Implementation and Enforcement of Environmental Law (IMPEL).
IMPEL-TFS. (2012). IMPEL-TFS Enforcement Actions III Project Report Enforcement of the European Waste Shipment Regulation Document control sheet Project: Client: Document title: IMPEL Enforcement Actions III IMPEL/SEPA Project No: Originated Document status, Brussels.
IMPEL-TFS. (2013). IMPEL TFS prosecutors project Final report, Brussel, Belgium.
Interpol. (2009). Electronic waste and organized crime. Assessing the links. Phase II report for the Interpol Pollution Crime Working Group. Interpol.
Iwama, T. (2004). Multilateral environmental institutions and coordinating reforms.
LNE, & Haskoning. (2010). *Onderzoek van de mogelijkheden van ketentoezicht op afvalstromen*. Mechelen: Haskoning Belgium NV—LNE Afdeling milieu-inspectie.
Nielsen, V. L., & Parker, C. (2009). Testing responsive regulation in regulatory enforcement. *Regulation & Governance, 3*(4), 376–399.
Parker, C., & Braithwaite, J. (2003). Regulation. In P. Cane & M. Tushnet (Eds.), *The Oxford handbook of legal studies* (pp. 119–145). Oxford: Oxford University Press.
Pellow, D. (2007). *Resisting global toxics: Transnational movements for Environmental Justice*, Cambridge: MIT Press.
Ruggiero, V., & South, N. (2010). Green criminology and dirty collar crime. *Critical Criminology, 18*, 251–262.
Sander, K., & Schilling, S. (2010). Transboundary shipment of waste electrical and electronic equipment/electronic scrap—Optimization of material flows and control. The Federal Environment Agency (Germany)–Ökopol GmbH.
Schluep, M., et al. (2011). *Where are WEEE in Africa. Findings from the Basel convention e-waste Africa programme*. Berlin: Secretariat of the Basel Convention, UNEP, EU, EMPA, Öko-Institut, IMPEL.

Sluis, A. van, et al. (2012). Nodal security in the ports of Rotterdam and Antwerp. In V. Bekkers & H. Fenger (Eds.), *Beyond fragmentation and interconnectivity: Public governance and the search for connective capacity* (pp. 73–94). Amsterdam: IOC Press.

Talley, W. (2009). *Port economics*. Florence: Routledge.

Van Daele, S., & Beken, T. V. (2009). Afvalcriminaliteit en kwetsbaarheden van de afvalsector [Waste crime and the vulnerabilities of the waste sector]. In G. Vande Walle & P. Van Calster (Eds.), *De criminologische kant van ondernemen [The criminological side of business]* (pp. 43–52). Den Haag: Boom Juridische Uitgevers.

Van Daele, S., Beken, T. V., & Dorn, N. (2007). Waste management and crime: Regulatory, business and product vulnerabilities. *Environmental Policy and Law, 37*(1), 34–38.

Van Erp, J., & Huisman, W. (2010). Smart regulation and enforcement of illegal disposal of electronic waste. *Criminology & Public Policy, 9*(3), 579–590.

Vander Beken, T. (2007). *The European waste industry and crime vulnerabilities*. Antwerp: Maklu Publishers.

Vig, N. J., & Faure, M. G. (2004). *Green giants? Environmental policies of the United States and the European Union*. Cambridge: MIT Press.

White, R. (2011). *Transnational environmental crime. Towards an eco-global criminology*. New York: Routledge.

Dr. Lieselot Bisschop is an assistant professor at John Jay College in New York City and postdoctoral research fellow of the Research Fund (FWO) at Ghent University (Belgium).

Chapter 4
Smuggling Networks and the Black Market in Ozone Depleting Substances

Lorraine Elliott

Introduction

The black market in ozone-depleting substances (ODS) is a direct consequence of international agreement on targets to reduce and phase-out the production and consumption of such chemicals. In effect, the Montreal Protocol on Substances that Deplete the Ozone Layer, which was agreed in 1987 and which came into force in 1989, created the opportunity structure for illegal activity in situations where perverse incentives generate high profits and low entry costs. Yet neither the Protocol nor the framework Vienna Convention, under whose auspices it was negotiated, say anything about illegal trade (as opposed to unauthorised trade) and it is widely accepted that the Parties had not anticipated such an outcome and were slow to respond.[1]

Since the Protocol came into effect, Meetings of the Parties (MOPs) have agreed a number of decisions on the topic of illegal trade. The language in these is generally cautious: no reference is made to criminal activity, primarily on the grounds that activities are seen to be defined as illegal and/or as criminal only in the context of the national legislation and regulations that Parties adopt to give effect to their obligations under the Protocol (see Montreal Protocol 2002, p. 2). The concern with illegal trade is driven by fears that it undermines the effectiveness of the Protocol and threatens Parties' compliance records, rather than any worry about the broader implications of transnational criminal activity. The Parties have directed most of their efforts to calls for tracking, labelling and licensing systems, better coordination and cooperation between relevant agencies, and more streamlined strategies for

[1] The Vienna Convention for the Protection of the Ozone Layer was adopted in 1985. Although it contained no specific targets for dealing with ODS, it did give effect to the precautionary principle which provides that scientific uncertainty should not be used as an excuse to avoid action in situations where the balance of evidence called for a political and legal response.

L. Elliott (✉)
Australian National University, Canberra, Australia
e-mail: Lorraine.Elliott@anu.edu.au

T. Wyatt (ed.), *Hazardous Waste and Pollution,* DOI 10.1007/978-3-319-18081-6_4

the exchange of information. The logic behind these moves is that by improving the management of legal trade, they will also help to prevent illegal trade even though the Parties have also recognised that such systems can still be evaded.

In theory, the black market in ODS is time bound: it will come to a natural end when all production and consumption of ODS is phased out under international law, through the development and adoption of substitute chemicals and as existing ODS-using equipment reaches the end of its useful life. In practice, the illegal trade has remained active over the last two decades even though it has involved different pollutants and engaged different geographies over that time.

This chapter examines this illegal trade. It provides an overview of the main components of the trade with a discussion of the chemicals involved and an analysis of the drivers of the trade and the patterns of illegal ozone supply chains. It concludes with some thoughts on the nature of smuggling networks and those involved in the illicit trade.

Defining the Illicit Market in ODS

The illegal trade in ODS is defined by the provisions of the Montreal Protocol and the steps that governments take to implement those provisions. The ozone-depleting (OD) chemicals that have been most prevalent in the illegal international trade are chlorofluorocarbons (CFCs), hydrochlorofluorocarbons (HCFCs) and halons. ODS for which there are limited volumes of production or for which there are very specialised markets and a narrow range of uses or a limited number of end users are less likely to be attractive to smugglers (see Montreal Protocol 2002, p. 6). CFCs and HCFCs, on the other hand, have had a wide range of uses particularly in refrigeration and in automobile air-conditioning units, as foaming agents and as cleaning solvents. Halons have a primary use in special purpose fire extinguishers and protection systems.

The procedures established under the Protocol and managed through MOPs, various working groups, and formal amendments and adjustments, identify controlled chemicals and set targets for phasing out the consumption and production of those chemicals.[2] Article 4 of the Protocol contains provisions to control trade, particularly that between Parties and non-Parties.[3] But it does not define what makes trade illegal. Indeed, the general approach under the Protocol is that illegality is identified when Parties adopt legislation and regulatory structures to implement control, licensing and phase-out provisions. In brief, industrialised countries were required to phase out domestic production and consumption of halons by 1994; CFCs, carbon tetrachloride and methyl chloroform entirely by January 1996; and methyl bromide

[2] The full list of chemicals subject to control provisions under the original Protocol and subsequent amendments includes CFCs, halons, other fully halogenated CFCs, carbon tetrachloride, methyl chloroform, HCFCs, hydrobromofluorocarbons, methyl bromide and bromochloromethane.

[3] Universal ratification of the Protocol and its amendments has rendered that provision moot.

by 2005.[4] The control provisions for HCFCs required a 90 % reduction in domestic production and consumption in developed countries by 1 January 2015. Under the Protocol, developing countries (known as the 'article 5 Parties') are subject to different freeze and reduction provisions. Control measures also allowed for export-related production in industrialised countries to meet the basic domestic needs of developing countries.[5] As explored later in this chapter, while this approach to differentiation facilitated developing country commitment to the Protocol, it also opened a significant loophole for illegal trade.

Market Drivers

The discussion later in this chapter reveals a continued market demand in both industrialised and developing countries for ODS for use in refrigeration, automobile air conditioning, and small business operations. That market has been exacerbated by price and profit differentials that have encouraged rather than dampened demand and that have attracted criminal entrepreneurs to this illegal commodity market. The complex control and trade provisions of the Protocol that have made production and consumption illegal in some parts of the world but legal in others have ensured a ready supply of ODS for diversion into black markets.

Price and Profit

To meet their obligations under the Montreal Protocol to freeze and reduce production and consumption of ODS, governments in the USA, the UK and in other parts of Europe adopted policies as early as the early 1990s to minimise demand and to encourage the development of substitutes. The combination of scarcity of legal supply and monopoly of production had the opposite effect and created an opportunity for illegal markets. For example, after January 1996, the only entities that could legally import ODS into the USA were those who held essential use exemptions, who had unexpended consumption allowances, or were authorised to destroy CFCs to remove them from the market. The US government also imposed a substantial tax on the sale of legal (stockpiled) CFC-12. This meant that as stockpiles of legal CFCs for recharging air-conditioning and refrigeration units became increasingly scarce, they also became increasingly expensive as did the costs of retrofitting units

[4] More detail on the control provisions under the Montreal Protocol can be found at http://ozone.unep.org/new_site/en/Treaties/treaties_decisions-hb.php?sec_id=6,7,8,9,10,11,12,13,14,15,16,17,18,19,20,21,22,23,24.

[5] Under the provisions of the Protocol, these 'domestic needs' provisions would last only until the date at which full phase-out for developing countries was reached. At that date, no Party can produce controlled substances except where there is provision for essential uses agreed under the Protocol.

to use CFC substitutes. Demand for cheap ODS therefore went 'underground', creating an opportunity for illegal supply and, equally important, illegal suppliers. The difference in price between legal and illegally sourced ODS also generated substantial profits for those involved in managing the illegal trade. In the mid-1990s, US officials calculated the profit ratio for smuggled CFCs at anything up to 13 to 1 (see Goldberg 1996).[6] Reports from the USA suggest that CFCs bought legally in Mexico for $ 150–185 a canister were being sold to auto shops in the USA for $ 300–450 (McArdle 2011) a profit of anything between 100 and 300 %.[7] In the early 2000s, CFC-12 purchased in China for US $ 1 a kilogram could fetch 12–16 times that on the UK black market (Newman 2001, p. 14). By 2011, HCFCs could be purchased in developing countries for little more than 2.00 € a kilo whereas the cost in the EU could be anything between 18 and 30 € a kilo (Environmental Investigation Agency 2011a). Similar price incentives and profit drivers characterised the black market in developing countries. In Thailand, for example, import duties of 30 % on ODS provided an incentive for smuggling activities to access cheaper sources. High domestic prices for ODS in India—a result of supply cartel activity—had a similar impact.

Differentiation and Loopholes

As noted above, the provisions of the Montreal Protocol adopted a differentiation approach to the production and consumption of ODS, providing less restrictive phase-out provisions for developing countries. This included allowing production in industrialised countries for export to meet the domestic consumption demands of developing countries. The logic for the former was to offer a grace period for developing countries in recognition of their limited technological and financial capacity for producing or adapting to the use of ODS substitutes. The logic for the latter was to ensure that production capacity was contained (admittedly mainly in industrialised countries) and therefore more easily managed for compliance with the provisions of the Protocol and domestic implementing legislation. Together these provisions created a ready supply for illegal exports, imports and trans-shipments. For example, the production of CFCs such as R-12/Freon was legal in Mexico, Russia, China and India for another 10 years after it became, in effect, illegal in industrialised countries in January 1996. Some of those production facilities in Mexico and India, as well as others in Brazil and Venezuela, were joint ventures with the major ODS producers such as US companies DuPont and Allied Signal and French company Elf-Atochem who together continued to control about 40 % of the global market. Chinese companies accounted for about the same proportion, with much of the rest produced in Russia and in India, where four companies were licensed to produce ODS (see Lobe 1997).

---

6 This compared with a ratio of only 4 to 1 for crack cocaine.

7 This was still far below the price of $ 850–1200 for the same quantity of legal but scarce CFC.

Other loopholes for illegal activity—and challenges for customs and border protection officials—were created by the complexity of the Protocol's trade and control provisions, the lack of an agreed labelling and naming system, and varying import and export licensing requirements that were legislated for at different times by individual Parties. The Protocol processes initially controlled only ODS chemicals and mixtures, but not products or equipment containing ODS. ODS for essential uses—either generally or for individual countries—were also exempt from control provisions until substitutes became easily and cheaply available.[8] It was not until 1999, when the Montreal Amendment to the Protocol came into force, that governments were required to put in place systems for licensing the import and export of controlled ODS including new, used, reclaimed and recycled chemicals. The wide range of names and labels used globally for ODS also created problems for enforcement agencies when it came to tracking and identifying the movement of ODS chemicals and products or equipment containing such chemicals. For example, the World Customs Organization's Harmonised Customs Code grouped chemicals such as HCFCs and halons together in broad categories rather than coding them individually.

Mapping the Illegal Market

The illegal trade was directed initially to meeting market demand for CFCs in developed countries, especially the USA and parts of Europe. The most commonly smuggled OD chemical in the 1990s was CFC-12, also known as R-12 or by the trade name Freon as marketed by DuPont Chemicals.[9] Freon was used mainly in the air-conditioning systems of cars manufactured prior to 1994. The destined end users for much of the ODS smuggled into the USA in the late 1990s and early 2000s were small automobile workshops and air-conditioning businesses who may or may not have known that they were purchasing illegal products. In the UK, demand came particularly from retailers and small businesses that required refrigeration systems.

As these markets began to shrink by the end of the twentieth century, in response to domestic regulation and increased enforcement, the illegal market shifted to other industrialised countries, particularly Japan. The demand for CFCs in Japan followed a pattern similar to that of the USA: over a quarter of the 70 million cars on the road were fitted with air-conditioning units that required CFCs to function (Clark 2003, p. 5). New markets for smuggled CFCs also opened up in developing countries in the face of international provisions to freeze consumption of ODS in such countries from 1999. According to Liu Ning (2007, p. xi), markets for illicit

[8] Essential uses included propellants for metred-dose inhalers and some agricultural, laboratory and analytical uses. They were either recognised as essential under the Protocol's procedures or nominated by Parties as essential in particular circumstances.

[9] R-12 was the designator given to CFC-12 by the American Society of Heating, Refrigeration and Air-Conditioning Engineers (ASHRAE).

ODS were expanding in China, India, Malaysia, Thailand, Indonesia, the DPRK (North Korea), the Republic of Korea (South Korea), Iran, Pakistan, Sri Lanka, Vietnam, the Philippines and Bangladesh. Demand in those countries remained high because of their continued reliance on equipment that used ODS and because substitute chemicals were expensive. So too was the cost of replacing OD-using equipment especially when that equipment continued to have a useful working life.

At the same time, a black market was also developing in the most common CFC substitute, HCFCs. HCFCs were intended to be a transition chemical on account of their lower OD potential but they were also always intended to be subject to phase-out provisions under the Montreal Protocol. As with CFCs, their uses have varied but they are predominantly used in air-conditioning and refrigeration systems.[10] While the European Union (EU) banned the import and use of virgin HCFCs from the beginning of January 2010, these chemicals were rapidly becoming the ODS of choice in many developing countries. By 2011, consumption was estimated to be growing by about 15 % per annum (Hurley 2011). In those parts of the world where bans on consumption and production were in force, demand continued to outstrip supply, thus creating scope for continued illegal production and trade. In 2012, US chemicals manufacturer DuPont estimated that in the USA alone, the demand for HCFC-22 in service industries would exceed 27.5 million pounds (weight) annually in the years to 2015.

While most of the interest in the ODS black market has focused on consignments of the chemicals in their gaseous form, this market has been further complicated by the development of an illegal trade in so-called 'banked ozone' held in products or equipment such as second hand refrigerators and chilling units that contain ODS. The trade in banked ozone began in the second half of the 1990s (see Dauvergne 2008, p. 109) although it was not controlled or made illegal under international law until the early 2000s. Reports from this period identify the growth of an illegal trade in products or equipment containing ODS. In 2000, the EU banned the export of used equipment containing ODS, thus potentially increasing rather than reducing the incentives for illegal trade. Indeed, companies in a number of European countries have been investigated for attempts at export of this kind in contravention of EU law. In 2007, customs authorities in Denmark twice stopped export containers that included used refrigerators containing CFCs. Two years before, eight companies had been prevented from similar illegal exports although only three were prosecuted (UNEP 2008, p. 4). In 2007, French authorities seized a number of CFC-charged refrigerators destined illegally for Senegal.

[10] The use of HCFCs has become a more general problem because they are also a greenhouse gas with a higher global warming potential than carbon dioxide.

Measuring the Trade

It is difficult to be precise about the size of the market. Detailed studies on ODS imports and exports have revealed significant irregularities, though illegal trade is only one possible explanation for these discrepancies.[11] Data on seizures provide one useful source of information although, as noted below, these data say little about what proportion of illegal trade is actually being intercepted. Seizures can range from small captures—individual cylinders carried by smugglers or concealed in cars—through to substantial consignments that represent repeat events over time. It is these larger and more regular smuggling efforts that have attracted most attention, in the main because the intention to deceive, conceal, and profit from illegal activity is clear.

From the middle of 1998–2003, US authorities seized 900 tonnes of CFCs although they estimated that as much as 7000 tonnes entered the USA illegally in that time (Clark 2003, p. 5). If the latter figure is accurate, authorities were intercepting less than 15 % of smuggled consignments. One of the largest cases prosecuted in the USA involved 418,654 kg of ODS—more than 29,000 cylinders in total—smuggled by the Kroy Corporation in 11 separate shipments between 2007 and 2009. Between July 2010 and August 2012, customs and enforcements officers involved in the Europe and Central Asia Enforcement Network reported seizures involving 3016 cylinders and 72 pieces of equipment containing more than 61 metric tonnes of ODS chemicals and mixtures (see UNEP DTIE 2012). Between September 2012 and April 2014, customs and enforcement officers in the same region reported 133 successful seizures of 7370 pieces of equipment (appliances and compressors) and 9513 refrigerant cylinders or containers which contained, in total, more than 467 metric tons of OD chemicals and mixtures (see UNEP 2015, p. 6–7).[12] Seizures have also been made at the point of export. In two separate incidents in 2010, French authorities intercepted the attempted illegal export of 44 tons of HCFCs (World Customs Organization 2010).

Two of the best known international operations on ODS have been the two phases of Project Sky-hole Patching coordinated by the United Nations Environment Programme (UNEP) and the World Customs Organization. In the first phase, which focused on the Asia Pacific, 21 customs agencies working collaboratively to share information and intelligence seized more than 700 tonnes of ODS between September 2006 and September 2009. The second phase expanded to include enforcement agencies in Europe, Africa and Central Asia and lasted from 3 May to 3 November 2010. More than 7500 cylinders of ODS (CFCs, HCFCs and others)—an

---

[11] Efforts to calculate the amount and value of illegal ODS are complicated also by the use of different methods of reporting and counting. Reports refer variously to the weight of contraband chemicals in pounds, kilograms, tons and tonnes (with the latter also being confused with metric equivalent ozone depleting potential tonnes). Other sources refer to the number of cylinders which come in multiple sizes.

[12] The countries involved were Albania, Belarus, Bosnia and Herzegovina, China, Cyprus, Georgia, Germany, India, Israel, Kyrgyzstan, Montenegro, Russia, Serbia, Spain, Tajikistan, Turkey, Ukraine and Uzbekistan, plus the EU.

estimated 108 tonnes in total—as well as 668 pieces of equipment containing ODS were seized and confiscated (see World Customs Organization 2010).

This snapshot of seizures indicates that the market is a global one. Indeed, information submitted to the Montreal Protocol Secretariat reports seizures in Antigua and Barbuda, Argentina, Armenia, Georgia, Kenya, Kyrgystan, Jordan, Nepal, Thailand, the Philippines, Serbia, the Seychelles, South Africa, Sri Lanka and Uzbekistan. The Environmental Investigation Agency (EIA) also reports seizures in South Africa, Antigua and Barbuda, and Jordan (Clark 2003, p. 4). A total of 74 tons of CFCs and HCFCs were seized in Nepal in 2001 and 2002 alone: the country's baseline quotas were 30 and 23 tons, respectively (Shandip 2004). Sri Lankan customs authorities detained 200 cylinders of Chinese-sourced CFC-12 in April 2003. In March 2004, authorities in the Republic of Georgia found illegal CFC-12, concealed in counterfeit cylinders and exported from the United Arab Emirates (almost certainly Dubai). In 2010, Thai customs confiscated 1200 cylinders of ODS. In 2012, 55 cylinders of R-12 were intercepted at the Kenyan port of Mombasa. As the discussion on logistic trails later in this chapter indicates, these are rarely isolated incidents.

This overview also gives some indication of how robust the illegal market might be. It tells us little about what proportion of smuggled chemicals and equipment is actually being intercepted. Nor does it give much indication of the overall value of that trade. Studies refer variously to illegal profits, to 'street value', to equivalent value on legal markets and to the cost to governments in lost excise and taxes. UNEP has estimated that in the mid-1990s the total amount of illegal CFCs in circulation was something between 16,000 and 30,000 tonnes, equivalent to 6–15 % of global production (Montreal Protocol 2002, p. 3). Educated best guesses put the value of this market at the time at around US$ 300 million annually (Brack 2001, p. 5). The amount smuggled into the USA alone was estimated at the equivalent of one-quarter to one-third of US production (Clapp 1999, p. 113). By 2000, as much as 15 % of CFCs used in the UK were thought to have entered the country through illegal means (Hayman and Trent n.d.). The amount being smuggled into India between early 1999 and March 2000 (more than 800 tonnes) was of a similar proportion, estimated as equivalent to 12 % of national consumption (IPS 2001). By the middle of the first decade of the 2000s, the illegal trade was thought to constitute between 7000 and 14,000 tonnes of ODS a year, estimated still at about 10–20 % of the legitimate trade (Clark 2007, p. 1).

Logistic Trails

The discussion thus far has identified the drivers of the black market in ODS and the overall patterns of trade associated with this market. It shows that as governments adopted more stringent regulations to meet their commitment to the Montreal Protocol and as the source of illicit ODS production shifted from developed to developing countries, so too did patterns of demand and the structure of the markets. This part of the chapter now turns to offer a more focused examination of how the actual trade in illegal ODS is managed and, where evidence is available, by whom. Tracing and analysing the logistic trails of illegal chains of custody in the ODS black market helps to map the nodes associated with production and export, consumption and import and, where relevant, transhipment. It also sheds light on strategies for concealment and methods of transportation.

Sources of Illegal ODS

As the discussion earlier in this chapter indicates, the manufacturing of ODS has always been limited to a small number of companies and countries. In the early phases of black market activity, much of the ODS entering the black market was being diverted from legitimate production sources, particularly those that were licensed to export ODS for domestic use in article 5 (developing) countries or to produce ODS for essential uses as determined under the Montreal Protocol. This, for example, was one of the sources of virgin (new) CFCs smuggled into the USA once production for domestic consumption was banned after January 1996. ODS that were exported legitimately from the USA to Mexico, an article 5 country, were then smuggled back into the USA to avoid excise and to take advantage of price differentials between increasingly scarce domestic stockpiled CFCs and black market sources. The North American Free Trade Agreement (NAFTA) Commission for Environmental Cooperation also reported CFCs being sold to Venezuela and Guatemala and then smuggled back into Mexico (Commission for Environmental Cooperation n.d.).[13] ODS for the illegal US market were also sourced from countries in which local companies, including those working in partnership with or as subsidiaries of global multinationals, were licensed to produce for domestic consumption needs in developing countries. Some reports suggest that by the mid-1990s much of the ODS smuggled into the USA was produced in India (Hanley 1996) and others show that CFCs were also being smuggled from Russia as well. In 1997, for example, UNEP estimated that Russia was producing 52,700 tonnes of CFC a year with seven factories the primary source of an estimated 30,000 tonnes on the black market (Land 1997).

The implementation over time of various phase-out schedules was accompanied by a shift in both production locations and in the chemicals produced and entering the black market. In its 2002 report for the Open-ended Working Group, the secretariat of the Montreal Protocol noted that of six developing countries with CFC production capacity, China was contributing 40 % of global production and India 20 % (Montreal Protocol 2002, p. 27). By the mid-2000s, India and the Republic of Korea had come to account for 70 % of global production of CFCs (Liu Ning 2007, p. xi). If consumption patterns are taken into account as well, the Asia Pacific accounted for over 80 % of the world's CFC production and consumption by the early 2000s.

[13] Mexico was not the only point of entry into the USA. Some illegal goods came in from or through Canada. Others came into USA ports such as Miami, Houston, San Ysidro and New York.

As a consequence, the region had also become a major hub for smuggled ODS, particularly CFCs and halons. India and China were the most prominent producers, consumers and transit points (Brack et al. 2006, p. 6), overtaking Russia in that regard. In 2002, China accounted for 90 % of developing country halon production (Montreal Protocol 2002, p. 27), and in 2012 it was reported as the source of more than half of the world's supply of HCFC-22 (Rosenthal and Lehren 2012). This also reflects the move from production of CFCs to production of HCFCs discussed above.[14] This move has been made easier by what Liu Ning (2007, p. 7) refers to as 'swing plants—facilities that can easily be converted' to produce related chemicals.

The phasing out of global consumption in specific ODS under the Montreal Protocol, including developing country consumption for domestic needs, had at least two consequences for the black market. First, it meant that chemicals increasingly entered the black market from recycled or reclaimed sources rather than from virgin sources. This included the growing black market in 'banked' ODS mentioned above. This also became a device for concealing the original sources of ODS. Reports to the Montreal Protocol Secretariat, for example, show ODS being traded or smuggled from or through countries that have no production facilities and often very little reclaiming or recycling capacity. The second consequence of more rigorous phase-out schedules was an increase in illegal production more generally. Production capacity in countries such as China and Russia has been diverted to the illegal production of banned chemicals. Delays in achieving full compliance with production restrictions have also meant a continuing and reliable source of ODS for the illegal markets. The decommissioning of Russia's production facilities, for example, was disrupted by political and economic turmoil. Evidence suggests that, following the closure of the last of six licensed CFC production facilities in China in 2007, at least some of that capacity, expertise and equipment was diverted into illegal production. At the end of 2014, China announced closure of five HCFC production lines. Yet this capacity constituted only a little over 15 % of the country's total HCFC production which will not be phased out completely until 2030 (see The World Bank 2014). There has also been a growing market in counterfeit ODS, first CFCs and more recently HCFCs. These are passed off as the product of companies such as DuPont, which still have legitimate production licences (for essential uses for example). They are sometimes sold openly as 'legal' commodities and sometimes diverted entirely into black markets.[15]

[14] Amendments to the Protocol in 2007 required developing countries to freeze HCFC production and consumption by 2013.

[15] DuPont has taken action against producers in China and traders in the United Arab Emirates, the latter linked to counterfeit HCFC-22 stock found in Egypt, Saudi Arabia and Jordan; see DuPont (2012a, b).

Concealment and Smuggling Routes

Information gleaned from seizure and other investigative reports shows that methods used to move illegal ODS across borders have involved almost all forms of transportation. Some smuggling efforts involve small amounts of ODS and one or two individuals only. Others involve larger amounts and more people but are still relatively simple to coordinate. For example, seizures have been made of individuals carrying cylinders strapped to their backs or transported on rafts, across the Rio Grande between Mexico and the USA, and across the Khong River between Lao PDR and Thailand, the latter possibly having come from China through Vietnam. Other interceptions have found individual smugglers transporting cylinders by cycle-rickshaw between Bangladesh and India (Liu Ning 2007, p. 20). Reports are also common of ODS concealed in cars, trucks or small boats that are often used for smuggling other goods as well. Polish authorities have found cylinders of CFC and HCFC smuggled by car from Ukraine on more than one occasion (UNEP 2008, p. 3). Similar interceptions have been made on the border between Serbia and Bulgaria. Uzbeki officials have reported regular confiscations of ODS from China (identified by markings) hidden in car boots: the goods appear to have been smuggled through Kyrgyzstan which shares a border with China. In April 2007, authorities in Kazakhstan found cylinders of R-22, R-409a and R-134a hidden in the ventilating shafts of a passenger train. Hurley (2011) reports activity in Southern Europe involving 'up to 10 small ships a day … ferrying small consignments of HCFCs between ports bordering the EU … and ports within the EU'. The Montreal Protocol Secretariat reported that 'tobacco boats' in the Caribbean, so called because that is what they were usually used to smuggle, were being seized and found full of CFC cylinders (Montreal Protocol 2002, p. 4).

The most common methods of illegal import and export involve false documentation and declarations—in effect, ODS are 'hidden in plain sight'. Shipping documents can be falsified to claim that goods are in transit to a legal market in a third country or that they are being imported for destruction. However rather than being shipped on, or being destroyed, chemicals are diverted into domestic markets. False documentation can also be used to assert that ODS are from reclaimed or recycled sources even though the source (export) country has no such facilities.[16] Other examples of fraudulent documentation include claims that containers are being returned empty to a source country when in fact they are full, or that they come from a single production source in cases where licensing authorities in the importing country have regulated against multisource imports.[17] Diversionary practices have also included mislabelling of containers to confuse customs officers by claiming

[16] Virgin substances have sometimes been adulterated to make them look as if they are reclaimed.

[17] One case in the USA, for example, involved sending large industrial refrigeration units to Venezuela to be refurbished before being returned. They were recharged with over 2500 pounds of CFC-12 even though they required only 3–4 pounds to work effectively.

that the contents are a non-ODS substance[18]; tampering with cylinder markings; and falsifying customs codes and chemical names. One of the largest cases in the EU, for example, involved 365 tonnes of halon-13-1 and 630 tonnes of CFC-12, fraudulently labelled as the non-OD chemical HFC-227.

At the other end of the spectrum, some illicit ODS chains of custody involve not only significant consignments in size but also the use of complicated methods of concealment. Indeed, as governments implemented more stringent control processes, methods of concealing the chemicals themselves, as well as their origin, have become more sophisticated and complex. These efforts have involved hiding illegal containers in cylinders that have been designed to meet pressure and temperature tests that would mark them as a legal ODS. These strategies can be used with double layering—hiding the illegal containers behind a row or two of legitimate substances. The routes used to smuggle ODS have also become more complex, involving a robust degree of organisation and conspiracy, the latter a term used often by the US Department of Justice in describing this particular black market. A 2012 case prosecuted in the USA, for example, involved ODS sourced in China, transported first to Ireland, and then to the Dominican Republic before being imported into the USA with fake invoices and falsified shipping documents (Rosenthal and Lehren 2012). The Bahamas and the Mexican port of Veracruz offered routes into the USA for ODS from China and Europe. Mexican manufactured ODS was reported to be entering the USA not only over the land border but also through the Virgin Islands. Seizures within the Asia Pacific have tended to come from regional sources, but UNEP reports at least one known case of Mexican CFCs ending up in Thailand and another where the import was seized from Egypt (Liu Ning 2007, p. 38, 49).

Transhipment hubs have also become important in the development of more sophisticated and wide-reaching smuggling networks and illicit chains of custody. The two key locations identified by governments and NGOs alike are Dubai in the United Arab Emirates and Singapore in Southeast Asia. Both offer free trade ports in which few, if any, commodity consignments destined for other markets are inspected. As Clark (2003, p. 1) points out, the shipment of consignments through transit countries 'confus[es] the trail of the material and provides a jump-off point into illegal markets'. Dubai has been a hub for ODS being transported to Pakistan and India as well as exported from illegal sources to other parts of the world. EIA reports have identified Singapore as a central node in illegal ODS markets where commodities are sometimes decanted and repackaged before being traded on to other destinations. Cambodia, Lao PDR and Vietnam have also become transit countries in Southeast Asia.

Other arrangements designed to facilitate all kinds of legal trade have facilitated illicit trade at the same time. For example, the port access agreements between Nepal and India have provided another convenient opportunity and route for smugglers. Those agreements guarantee the landlocked country of Nepal free access to

[18] HCFC-22, for example, is often deliberately mislabelled as R-134a, a non-ODS hydrofluorocarbon that is used in new and retrofitted car air-conditioning systems.

Indian ports such as Kolkata (Calcutta) for export and import purposes. ODS have been brought into Nepal from the EU under the provisions allowing for export to article 5 countries (though much in excess of Nepal's consumption quotas) and then smuggled back into India across land borders for use in refrigeration and air conditioning there.

Conclusion: Entrepreneurial Networks

As with the legal trade in ODS, and as the discussion in this chapter has revealed, illicit chains of custody can involve networks of manufacturers, processors, distributors and traders with 'several links in the chain between producer and end user' (Montreal Protocol 2002, p. 9). The market ranges across 'small-time entrepreneurs to sophisticated money-laundering conspiracies' (Baker 2000). Various reports on the illegal trade refer to 'extensive web[s] of middlemen and brokers' (Newman 2000, p. 3037), 'notorious smugglers' (EIA 2011b), 'dishonest brokers' (Clark 2003, p. 4), 'sophisticated white collar crim[inals]' (Zagaris 1996) and the 'ozone mafia' (Jernelov 2005, p. 9). The fact that the illegal trade has continued over time with little apparent disruption suggests that knowledge about smuggling networks has been widely known among the relevant community and that those involved, whether new players to the market or not, were able to adjust quickly to adapt the various stages of chains of custody to new sources and new buyers. In this context, then, illicit ozone networks and chains of custody constitute what Passas (2002, p. 16) refers to as entrepreneurial structures through which criminal acts—in this case the various functions that support the trade in illegally sourced or produced ODS—are carried out.

In some cases, those entrepreneurial structures take the form of the kind of 'hub' (or star or wheel) network that Arquilla and Ronfeldt (2001) characterise as a form of cartel or franchise. In such networks, they suggest, actors are linked to a central (though not necessarily hierarchical) node through which coordination is managed (Arquilla and Ronfeldt 2001, p. 7). This seems to have been the pattern behind a number of cases in the USA where those organising the smuggling were often associated with or had links to legitimate companies and enterprise that could be used as a front for import or a distribution point for resale. The individuals at the hub of such networks were rarely working alone to coordinate illegal consignments. Importing companies—or their principals—have been found to be working closely with distributors and suppliers (see, for example, EPA 1999). Those who manage these networks and who arrange the purchase of illegal sources of ODS are known to have provided advice to importers, to express-mail fake invoices to exporters in other countries, to have used aliases and to have established front companies and offshore bank accounts in fictitious names.

Other illicit chains of custody in the ODS black market have taken a form that is more akin to a chain network, a flatter form of entrepreneurial structure in which people, goods, or information move along a line of separated contacts, and where end-to-end communication must travel through the intermediate nodes (Arquilla and Ronfeldt 2001, p. 7). ODS smuggling of this kind is likely to reflect a clear division of labour between those who do the physical transportation and those who are the buyers and distributors, with middlemen or brokers involved so that there is no direct link between producers and distributors. The case of the Alghazoulis brothers in the USA, for example, shows that they bought from a number of 'independent' suppliers who were bringing small consignments in from Mexico on a weekly basis and who were probably smuggling for others as well (see McArdle 2011). Other reports show similar patterns, with Mexican smugglers confirming that they would transport consignments to Houston to a middleman who paid them in cash (Sheff 1997).

In sum, then, three conclusions can be drawn from the narrative and analysis in this chapter. First, ODS smuggling now has a global reach. Second, black markets in ODS blur the boundaries between legitimate industry on the one hand and white collar and corporate crime on the other. Third, network structures have made it possible for those involved to adjust their strategies for illegal transboundary activity in response to changing structures of demand and supply.

References

Arquilla, J., & Ronfeldt, D. (2001). The advent of netwar (revisited). In J. Arquilla & D. Ronfeldt (Eds.), *Networks and netwars: The future of terror, crime and militancy*. Santa Monica: RAND.

Baker, L. (2000). Illegal CFC smuggling destroying the ozone layer, *EMagazine*. http://www.alternet.org/print/story/10109/illegal_cfc_smuggling_destroying_ozone_layer. Accessed 14 Feb 2015.

Brack, D. (2001). The scope of the problem: An overview of the illegal ODS trade. *OzonAction Newsletter, 6*, 4–7.

Brack, D., Saunders, J., Tarasofsky, R., Bass, C., von Bismarck, A., Clark, E., & Grabiel, D. (2006) *ODS tracking: Feasibility study on developing a system for monitoring the transboundary movement of controlled ozone-depleting substances between the parties*. London: Chatham House/ Environmental Investigation Agency.

Clapp, J. (1999). The illicit trade in hazardous wastes and CFCs: International responses to environmental "bads". In H. R. Friman & P. Andreas (Eds.), *The illicit global economy and state power*. Lanham: Rowman & Littlefield.

Clark, E. (2003). *Lost in transit: Global CFC smuggling trends and the need for a faster phase-out*. London: EIA.

Clark, E. (2007). *Ozone depleting substances—the global illegal trade: History and current trends*. Paper presented at the Meeting of the Regional Ozone Network for Europe and Central Asia, Ashgabat, Turkmenistan, 26 February to 2 March.

Commission for Environmental Cooperation. (n.d.). The illegal trade in chemicals that destroy ozone, *CEC Update*, Spring. www.cec.org/Page.asp?PageID=122&contentID=2491&StieNodeID=457.

Dauvergne, P. (2008). *The shadows of consumption*. Cambridge: MIT Press.

DuPont. (19. December 2012a). DuPont completes successful action against counterfeiting in Quzhou City, China. *News Release*.

DuPont. (2. July 2012b). DuPont completes successful action against counterfeiting in United Arab Emirates. *News Release*.

Environmental Investigation Agency. (16. September 2011a). HCFC phase-out at risk from illegal trade. *Press Release*.

Environmental Investigation Agency. (2011b). Public and private sectors in partnership to face new challenges with trade in ozone depleting chemicals, 17 June. http://eia-international.org/public-and-private-sectors-in-partnership-to-face-new-challenges-with-trade-in-ozone-depleting-chemicals. Accessed 14 Feb 2015.

Environmental Protection Agency (USA). (29. January 1999). Californian man pleads guilty in CFC smuggling conspiracy. *News Release*.

Goldberg, C. (10. November 1996). A chilling change in the contraband being seized at borders. *The New York Times*. http://www.nytimes.com/1996/11/10/us/a-chilling-change-in-the-contraband-being-seized-at-borders.html. Accessed 14 Feb 2015.

Hanley, C. J. (28. March 1996). Smuggling of Freon gas into the U.S. threatens recovery of ozone layer. *Deseret News*. http://www.deseretnews.com/article/print.

Hayman, G., & Trent, S. (n.d.). New challenges to the Montreal Protocol. Environmental Investigation Agency, London. http://www.unesco.org/education/educprog/ste/newsletter/eng_n2/newchall.html. Accessed 14 Feb 2015.

Hurley, N. (1. July 2011). Land of the blue sky, Mongolia hosts meeting on the illegal trade that threatens our ozone layer, *EIA Investigator's Blog*. http://eia-international.org/mongolia-ozone-depleting-substances-un. Accessed 14 Feb 2015.

IPS-InterPress Service. (14. October 2001). Anti-ozone chemicals enter … New Delhi. *Press Release*. http://ban.org/ban_news/anti_ozone.html. Accessed 14 Feb 2015.

Jernelov, A. (10. May 2005). The rise of the ozone mafia. *The Taipei Times*, p. 9.

Land, T. (1997). The greening of the black market? http://www.icis.com/resources/news/1997/01/07/44834/the-greening-of-the-black-market-/. Accessed 14 Feb 2015.

Liu Ning. (2007). *Illegal trade in ozone depleting chemicals*. Bangkok: UNEP.

Lobe, J. (1997). U.S. moves against CFC smugglers. *Albion Monitor/News*. http://www.monitor.net/monitor/97016/cfcarrest.html.

McArdle, J. (2011). Three Freon-smuggling brothers, a lingering mystery. *EENews*. http://www.eenews.net/stories/1059956197/print. Accessed 14 Feb 2015.

Montreal Protocol. (2002). Open-Ended Working Group of the Parties to the Montreal Protocol. *Study on monitoring of international trade and prevention of illegal trade in ozone-depleting substances, mixtures and products containing ozone-depleting substances (Decision XIII/12)*, UNEP/OzL.Pro/WG.1/22/4, 23 April.

Newman, J. (2000). New directions: Uncovering the illegal trade in CFCs and halons. *Atmospheric Environment, 34,* 3037–3038.

Newman, J. (2001). The tricks of the trade: How criminals smuggle ODS. *OzonAction Newsletter, 6,* 14.

Passas, N. (2002). Cross-border crime and the interface between legal and illegal actors. In P. C. van Duyne, K. von Lampe, & N. Passas (Eds.), *Upperworld and underworld in cross-border crime*. Nijmegen: Wolf Legal Publishers.

Rosenthal, E., & Lehren, A. (7. September 2012). As coolant is phased out, smugglers reap large profits. *New York Times*.

Shandip, K. C. (19. September 2004). Open borders hamper Nepal's commitment to phase out ODS. *The Kathmandu Post*.

Sheff, D. (August 1997). The chilling effect. *Outside Magazine*. http://www.outsideonline.com/outdoor-adventure/The-Chilling-Effect.

The World Bank. (15. September 2014). China announces major reduction in ozone depleting gases on International Ozone Day, *Press Release*. http://www.worldbank.org/en/news/press-release/2014/09/15/china-announces-major-reduction-in-ozone-depleting-gases-on-international-ozone-day. Accessed 14 Feb 2015.

United Nations Environment Programme (UNEP). (2008). *Information reported by the Parties to the Montreal Protocol on illegal trade in ozone-depleting substances (paragraph 7 of decision XIV/7)*, UNEP/OzL.Pro.20/INF/3, 7 October.

UNEP. (15. January 2015). Europe and Central Asia. *OzoNews, XV,* 6–7.

UNEP Division on Technology, Industry and Economics [DTIE]. (12. October 2012). ECA customs cooperation meeting on ozone-depleting substances and ECA ozone protection award for customs and enforcement officers in Istanbul, Turkey, 9–11 October 2012. *Press Release*.
World Customs Organization. (10. November 2010). Joint WCO/UNEP Operation nets large haul of ozone depleting substances, *Press Release*. http://www.wcoomd.org/en/media/newsroom/2010/november/joint-wcounep-operation-nets-large-haul-of-ozone-depleting-substances.aspx. Accessed 14 Feb 2015.
Zagaris, B. (1996). Miami person pleads guilty in [sic] smuggling CFCs after arrest in and extradition from Costa Rica. *International Enforcement Law Reporter, 12*(9), 363–364.

Professor Lorraine Elliott is a Professor of International Relations and ANU Public Policy fellow at the Australian National University in Canberra, Australia.

Part II
Environmental Justice Concerns

Chapter 5
Pollution, Access, and Binary Division: Water Activism and a Human Right to Water

Bill McClanahan

Water Access and Pollution as Binary Concepts

Water pollution, whether from point sources such as 2010's Deepwater Horizon oil spill or nonpoint sources such as quotidian stormwater runoff, constitutes one of the most pressing issues of global ecological health faced today (see White and Heckenberg 2014, p. 158; see generally Brisman 2002). Meanwhile, inadequate access to safe and sanitary supply of fresh water causes over 3 % of all human deaths worldwide—and is the leading cause of death among children under 5 years—while freshwater animal species face an extinction rate five times that of terrestrial animals (McClanahan et al. In Press; Prüss-Üstün et al. 2008). Although developing nations bear the brunt of insufficient access to clean water (see, e.g. La Franiere 2006), problems of accessibility are less likely to impact the developed and postindustrial world, unless they affect agricultural production or recreation: Droughts that crippled Midwestern US farmers and the closure of unsustainable golf courses each garnered significant media attention in 2012.

Within criminological and public imaginations, there seems to be a tendency to conceptualize relevant water concerns along the binary lines of access issues and pollution issues. For the most part, we have considered issues of water pollution as primarily affecting the first-world nations of the Global North (e.g. the 2013 Deepwater Horizon well explosion or the 1989 Exxon Valdez disaster), while issues of access are understood as primarily relevant to the developing nations of the Global South (e.g. movements for indigenous control of water supply systems in Bolivia and Ecuador) (McClanahan 2014; McClanahan et al. In Press). This tendency to divert water issues into one of these two streams has left us blind to the increasing likelihood that, with the global adoption of neoliberal logics advocating access-reducing water privatization schemes coupled with the growing effects of climate change and (frequently related) increased pollution events (Block 2013), the lines between issues of access and pollution will become less clear. This blurring will

B. McClanahan (✉)
University of Essex, Colchester, UK
e-mail: wwmccl@essex.ac.uk

T. Wyatt (ed.), *Hazardous Waste and Pollution,* DOI 10.1007/978-3-319-18081-6_5

likely occur not only along conceptual lines but also material ones, as those in the Global North face increased water shortages, while those in the Global South face increased pollution. In this chapter, I will explore the connections between access and pollution, various activist responses to issues of both access and pollution, and the potential of activist perspectives to highlight the increasing global interconnectedness of water issues.

Bridging the gap and blurring the geopolitical and spatial boundaries between issues of access and pollution are global claims of water as a human right. These claims, often coming from water activists, challenge various elements of an ever-expanding neoliberal global capitalism that contributes to both pollution and decreased access. If green criminology is to develop an understanding of the multitude of water issues at play in the world, it must be willing to dismantle some of the assumptions about the geography of those issues and then work to globalize conceptualizations of water issues (McClanahan et al. In Press). Key to a criminological understanding of water issues on a global scale is a willingness to understand activists concerned with water issues as working toward a common goal: water as a fundamental human right. In addition, I argue that both privatization and pollution must be understood as flowing forth from the global forces of neoliberal capitalism, and that such an understanding can aid in the development of a critical green criminology of water.

(Re) Emerging Issues of Water Access and Water Conflict in the Global North

In the contemporary social memory of the Global North, the 1989 Exxon Valdez disaster, the 2010 Deepwater Horizon well explosion and blowout, and the 2014 poisoning of Virginia's Elk River are perhaps the most salient images of 'brown crimes' (White 2005). Each of these events captured the public attention to varying degrees[1], each constitutes a major harm to water, and each can be understood broadly as involving some causal element of criminality. Following the Exxon Valdez and Deepwater Horizon disasters, popular media inundated the public with images of oil-soaked birds, fish and beaches, and these images appear to have had a lasting and significant impact on the social memory of the Global North (Joling 2014; Taylor 2014), with several media outlets running stories at various anniversaries of the spill describing the images as 'haunting' (Aguilar 2014; Strauss 1999), 'heartbreaking' (Atkin 2014), 'something we just can't forget' (Hseih 2014), and

[1] A cursory search of *The New York Times* archive suggests that over 1000 articles relating directly to the Deepwater Horizon oil spill have appeared in that publication alone between April 2010 and July 2014. Similarly, a search for terms relating to the Elk River chemical spill shows over 100 articles published in *The New York Times* alone in the 7 months between the disaster and this writing. Note, however, that some residents of Charleston have indicated that the Elk River chemical spill did *not* receive sufficient media attention. Osnos (2014:47) reports "In Charleston, people told me that their ordeal had received less national coverage than the latest virus on a cruise liner". For an analysis of media coverage of the Deepwater Horizon oil spill, see McClanahan et al. (In Press), Paulson et al. (In Press).

even 'cliché' (Atkin 2014). In the case of the Elk River event, wherein the river was contaminated when a local industrial actor, Freedom Industries, involved in the regionally dominant coal industry, allowed large containers storing a variety of hazardous chemicals to leak into the river, national media in America employed images of dead fish and local residents were forced to rely on bottled water for all their water needs. The thread connecting these events in the criminological imagination has, thus far, been a broad concern over regulatory inefficacy or malfeasance (for a general discussion of regulatory issues, see Du Rees 2001; South et al. 2014; Stretesky et al. 2013). What connects these events in the public imagination, however, is a basic feeling of concern over water pollution, and it seems that this concern has, in many ways in the Global North, trumped concerns over water access.[2]

Contrary to the ease with which pollution has captured the public imagination of the Global North, issues of access to water for consumptive, recreational, and agricultural purposes are being revealed with increasing urgency and regularity[3]. States in the western region of the USA, for example, have faced tremendously crippling droughts throughout 2013 and 2014 (see, e.g. Associated Press 2014a; Bostok and Quealy 2014; Burke 2014; Dwyer 2014; Murphy 2014; Powell 2014; Reid 2014; Smith 2014; Walker 2014; Woody 2014a, b). Several states have implemented new—or tightened existing—water-use limits for home consumers, and have cracked down on violators with increased fines and water shutoffs (see, e.g. Associated Press 2014a, b; Barnett 2014; Powell 2014; Steinmentz 2014). These regulations have, in many ways, revealed the gaps in economic and social capital that govern the accessibility of water by revealing the ability of wealthy and powerful water consumers to avoid use restrictions and regulation altogether; California's agricultural users who accountfor over 80 % of total water consumption in the state (Waldie 2014), are frequently exempted from use restrictions owing to both an outdated supply system incapable of measuring large-scale agricultural use and a continued reliance on arcane water rights developed in the initial Western migration of human populations (Los Angeles Times 2014). Similarly, residential access to clean water has been severely impacted in areas of the Midwestern United States such as Des Moines, Iowa, where large agricultural actors such as mega farms and concentrated animal feeding operations (CAFOs) have significantly affected public water supplies. These industrial agricultural operations have added such high levels of nitrates to local water sources through the use of fertilizer and the production of animal waste that nonagricultural users have seen water prices rise, as the

[2] As discussed further below, some pollution concerns are closely related to access concerns, as in the Elk River case; because the river supplied local residents with their water, pollution was understood to directly impact access. In the cases of the Deepwater Horizon disaster and the Exxon Valdez spill, however, pollution remained the primary concern, seemingly because significantly large populations did not rely directly on the affected bodies of water for their daily water needs (although local economies and individuals do rely on them for recreational, agricultural and industrial use).

[3] It should be noted that while concerns over water access had previously waned in the region of the USA recently affected by drought, the western states had, during their initial settlement and development, significant conflicts relating to water access. These conflicts, though, were for the most part settled by the early twentieth century. For a general discussion of those earlier conflicts, see Reisner (1993), Piper (2006).

local provider, Des Moines Water Works, is forced to employ a massively expensive treatment system in order to restore water purity to safe levels (Cosier 2014). In a time of sweeping drought in the USA, the ability of agricultural water users to pollute or consume huge amounts of water highlights both the emergence of access issues in the Global North and the intersection of access and pollution issues; pollution can often lead to increasingly restricted access.

Wealthy nonagricultural landowners in the USA have also found ways to outmanoeuvre usage restrictions: Oprah Winfrey, the wealthiest woman in the USA, owns several homes in Southern California, and has begun to truck in water to ensure that the ample and lush lawns and grounds of her property stay that way, at an annual cost of over USD $ 125,000. California's enforcement of water use restrictions has resulted in an influx of revenue to the state in the form of fines levied against residential consumers (the Montecito Water District, provider to Oprah and several other wealthy consumers, admits to receiving over USD $ 4 million in fine payments for 2014 alone), while massive agribusiness operations and wealthy residents enjoy essentially unfettered access to water without facing any legal or economic penalties (Bardach 2014; Walker 2014). The ongoing drought in the western USA has also contributed to the creation of a black market for water. 'Water poachers' are rapidly becoming a newly defined category of criminal offender, with community bonds tested not only by drought but also by water theft, or 'illegal diversions' (Bernard 2014). Prosecution for diversion of water is hardly new—the case of Gary Harrington in Oregon resulted in heavy fines and a prison sentence for Harrington, who was convicted of diverting rainwater and snow melt into his private holding ponds (McClanahan 2014)—but thefts of water in the drought-stricken states of the West are increasingly from private holding tanks filled with water purchased from municipal sources. The difference, then, between this emerging crime of water poaching and the sort of diversion Harrington was charged with is that while Harrington 'stole' water that was claimed by the state using a tenuous legal argument relying on old legislation, the new 'water poachers' are taking water from their neighbours. Harrington, it could be argued, claimed water that belonged to the commons, while the new water poachers take water that 'belongs' to their neighbours. The emerging category of water poaching, then, causes conflicts between individual actors, rather than conflict between the individual and the state. While there is a valid argument to be made against the state's claims of ownership of water based on the conceptual difficulties in understanding claims of ownership over something as essential and of the commons as water (see McClanahan 2014), claims of ownership become less conceptually difficult when made by an individual who has paid for water.

Drought, though, is not the only factor that highlights—or has the potential to highlight—issues of water access and concurrent conflict in the developed nations of the Global North. Social and material capital is a major determinant of individual and regional access to useable water supplies. This has been made clear by a number of recent incidents and issues relating to access and privatization in the Global North, with Detroit, Michigan's ongoing water troubles being the most revelatory and striking. Unlike the American West, Detroit has not suffered from an extreme degree of drought recently (US Drought Monitor 2014). Average to high levels of rainfall, however, have not stopped segments of Detroit's population from suffering from issues of reduced water access owing, in part, to the pervasive logic of water privatization. Here, in a city with an economy and social structure already decimated by the rapid loss of industry in the 1970s and 1980s, restricted water access stems from poverty and the unmeetable demands of a water supply system reflective of the logics animating neoliberal capitalism. While the image of the thirsty poor might be most salient in the public imagination when it reflects binary assumptions of abundance in the Global North and scarcity in the Global South, as suggested earlier in this chapter, these images and assumptions fail to reflect the growing gap in water wealth in the USA. For Detroit, this gap is created by a municipal water system that, according to activists, is attempting to leverage its debts—which total an estimated USD $ 5 billion—against continued public ownership, arguing that if it is to remain able to supply the area with water, privatization is necessary (Abbey-Lambertz 2014). Issues of water access—and activist responses to those issues—in areas of the Global North like Detroit invite a critical rethinking of the pollution-access binary by highlighting international activist concerns over access. While water access is an emerging issue in parts of the Global North, it has long been an issue confronted by activists and communities in the Global South. Similarly, while access to water has long been a concern for the Global South, the forces of globalization and the growing demand for energy and energy security for the Global North have brought increased pollution to the waters of the Global South.

Emerging Issues of Water Pollution in the Global South

In 1998, the government of Bolivia, under the supervision of the World Bank, passed laws that effectively privatized the water supply system of Cochabamba, a region that over 1 million Bolivians call home, 60 % of whom are indigenous. Concessions to manage—and profit from—the water system were granted to Bechtel, a multibillion dollar global engineering and construction corporation. Bechtel, upon taking control of the Cochabamba water supply, immediately tripled prices and cut off water services to those unable to pay, going so far as to charge for rainwater gathered in homemade and traditional catchment systems. In response, the Coalition in Defense of Water and Life, was formed, quickly organizing a referendum that demanded the cancellation of all Bolivian water contracts with Bechtel. When the Bolivian authorities refused to back down, protestors took to the streets and were met with repressive violence from the state military, leaving dozens wounded and a 17-year-old protestor dead. The Bolivian government eventually capitulated, severing all contracts with Bechtel (Barlow 2008a, pp. 102–106; McClanahan 2014; McClanahan et al. In Press).

Similar issues have arisen across the Global South: Ecuador, India, Nigeria, Vietnam, and Mexico have all seen water privatization promoted by international lenders as a way to increase the marketization and commodification of resources and spaces formerly conceptualized as part of the commons, and each has seen activist movements of varying size and efficacy resisting privatization (Barlow 2008b). The compelling images of drought, thirst, and access-based resistance movements—particularly those in Bolivia, which garnered significant media attention during election campaigns for president Evo Morales, who supported indigenous movements resisting privatization—has established the primacy of access issues (which frequently stem from privatization) in the Global South. Just as emerging issues in the Global North reveal the importance of access there, however, emerging concerns over pollution in the Global South reveal that issue's growing salience.

In the Niger River Delta, for example water pollution stemming from oil exploration and extraction by Western corporate interests like Shell has resulted in a massively polluted waterscape (see Nurse, this volume). The activities of corporate actors—including three major oil spills between 2004 and 2007—in the region has created myriad problems for the land, water, animals, and people of the Niger Delta, including a reduced life expectancy, 'pollution of much of the region's vegetation, fishponds and drinking water', and significant reductions to local social health (Nurse 2013). The situation in the Niger Delta has made its way into the public and criminological imaginations, in part, through increased media attention in the Global North (between 2000 and 2014, *The New York Times* alone, for example ran 24 stories on pollution in the Niger Delta), but also through academic attention (see generally: Nurse, this volume; Omotola 2009). Along with similarly polluted water in other areas of the Global South, including Indonesia (Greenpeace 2013), Malaysia (Mok 2014), Mexico (Gonzalez Covarrubias 2014) and South Africa (Crowley 2014), water quality conditions in the Niger Delta reveal that while water concerns in the Global South are frequently conceptualized around issues of water access and privatization, pollution is a rapidly emerging issue.

Pollution, Access, and Bottled Water Marketing

While drought-prone areas in the Global North now regularly enact legislation to limit water use, the majority of Americans still have giant conceptual hurdles to leap in understanding and responding to water scarcity; water scarcity, in the context of an America overflowing with markets selling bottled water, seems almost ridiculous. The false perception of abundance and unfettered access fostered by the availability and ubiquity of bottled water, however, ultimately contributes to problematic practices relating to issues of both access and pollution. As Brisman and South (2013, 2014; see also South and Walters 2014) have noted, the selling of bottled water involves the construction and perpetuation of various myths relating to access to clean water. Consumers within the Global North are encouraged, through the languages of marketing and conspicuous consumption, to distrust the water that flows from their taps—water that costs a fraction, per litre, of bottled water. Water sellers incorporate imagery tailored to evoke traditional notions of water purity—the glacier, the mountain stream, the natural aquifer—and promote the image of bottled water as an ethical, 'green', or health-conscious alternative to tap water, while, through stringent testing standards, municipal tap water supply systems maintain purity levels significantly higher than those of their bottled competitors. These myths touch the core of American water consciousness in that they appeal to a public increasingly inundated with—and concerned about—events involving water pollution; the Deepwater well disaster, the discovery of large amounts of pharmaceutical compounds in public water supplies in 2008 (Donn et al. 2008), and the rupture of an Exxon oil pipeline in Arkansas in 2013 (Caplan-Bricker 2013) have each garnered significant media attention focused primarily on how these events compromise water quality and the health of marine ecosystems. Indeed, the Exxon Valdez spill in 1989 still occupies a place in the public imagination and vernacular as a truly catastrophic event, more than 25 years later.

The willingness—even eagerness—of the public to consider seriously these harmful events illustrates not only the tendency, within the Global North, to view pollution as the most pressing issue relating to water but also that the fear of pollution is simultaneously a fear of reduced access. Here, again, it becomes clear that while pollution and access issues are often understood as separate issues with different geopolitical and spatial boundaries, they are increasingly a single issue. While instances of water pollution—still fresh in the American imagination—that are employed discursively by bottled water sellers are, of course, an issue of exceptional importance, this tendency within the developed world to give primacy to concerns of water pollution highlights what we might be missing by failing to consider equally important and perhaps inseparable issues of water access. Indeed, the reality that water sellers try so stringently to obfuscate is that issues of water access and water pollution are increasingly becoming inextricably linked. The catchment, manufacture, transportation, and sale of bottled water—those energy-intensive processes that ultimately coalesce to give the developed world such a mistaken perception of water abundance and easy access—is a significant contributor to the global degradation of water quality; the industrial processes of a manufactured and illusory abundance of water actually contribute significantly to reduced access to dwindling sources of fresh water (Food and Water Watch n.d.). Furthermore, as global warming continues apace, it is likely that some regions already affected by drought will be hit even harder, while others may find themselves rich in water (Agnew 2012; Brisman 2013). The constantly shifting geographies of water wealth and drought, combined with the increased likelihood of water pollution made possible by the opening of Arctic seas for shipping and oil exploration (Morin 2013) and the continuation of harmful consumptive practices, makes it ever more likely that the boundaries—conceptual and spatial—between access and pollution will blur further. This blurring will make the exchange of technologies, methods, and logics developed to reduce pollution and combat privatization an imperative for those wishing to ensure access to clean water, as well as those wishing to reduce water pollution. With an eye toward the interconnectedness of these issues, I will now explore various responses to issues of access and pollution.

Activist Responses to Issues of Pollution and Access

In the Global South, activist movements surrounding issues of privatization—and, thus, issues of access—have been underway for quite some time. Bolivia, as mentioned above, has been the site of frequent clashes between various governmental, corporate, economic, and activist concerns. In Chile, activists have made significant efforts to protect glaciers, a primary source of fresh water, from depletion at the hands of extractive industry (Torres-Leclercq 2014). India, particularly the Yamuna and Ganga river regions, has also seen significant mobilization of activist efforts to slow or halt privatization of water supply systems (One World South Asia 2006). Because a central concern surrounding water privatization is the reduction of local access due to increased costs of provision, these movements highlight, first and foremost, issues of water access.

Meanwhile, responses to water issues in the Global North continue to highlight the primacy, within the public imagination, of pollution. Concerned about the environmentally deleterious effects of wasteful and polluting American water habitus, green-minded North American water activists advocate changes in technology and practice that aim to correct harmful errors in human interactions with the water cycle. Recognizing the ease with which legal and regulatory measures can be skirted by the powerful—as evidenced by the aforementioned Winfrey's ability to maintain high levels of water consumption in the face of extreme drought—water activists in the Global North often focus their efforts on 'do-it-yourself' (DIY) approaches to water purity and access. These DIY approaches, methods, and technologies include the implementation of greywater systems designed to reuse 'wastewater', composting toilets designed to both conserve water and limit pollution, and homemade catchment and holding systems designed to limit reliance on municipal and private water providers (McClanahan 2014).

Of particular importance in the domestic fight is the California group Greywater Guerrillas. Formed in the late 1990s, the group advocated, designed, and installed greywater systems in what was often open defiance of municipal codes and regulations. Most compelling, within the context of green criminology, are the ideologies and logics that undergirded the quiet resistance of the greywater guerrillas. Adopting a tongue-in-cheek militancy—reflected in the name 'Greywater Guerrillas' which was later changed to Greywater Action in an effort to stop the possible alienation of potential allies in regions where 'guerrilla' carries with it unpleasant implications of violence (Allen and Woelfe-Erskine n.d.)—the group constructed a mythologizing image of the domestic water warrior, installing simple DIY greywater systems under the cover of darkness, armed with copper pipe, fittings, and plumbers' putty. By giving new meaning to Edward Abbey's radical 'monkeywrenching', the Greywater Guerrillas and similar 'plumbing advocacy' groups situate politically resistant DIY water reuse and catchment systems within a larger context and history of ecologically minded resistance to neoliberal privatization schemes.

The systems advocated by the Greywater Guerrillas were simple, low-tech affairs designed to control the flow of wastewater cleanly and simply. In 2007's *Dam*

Nation: Dispatches from the Water Underground—a title that again reflects the alignment the group felt with other militant resistance groups—the group produced a text that is part polemic, part how-to manual. The specific systems described in the book are simple, mostly consisting of diversion valves installed in traditional runoff pipes that allow the user to divert the flow of wastewater to reuse it for gardening, cleaning, and myriad other household tasks that do not require 'clean' water. Greywater systems are, in the context of the North, used primarily to reduce the pollution impact of household water use on municipal water supplies; running throughout the literature produced and disseminated by greywater activists and groups in the Global North is a focus on reducing pollution. This focus further illustrates the primacy of pollution in conceptualizations of water issues in the regions of the North. Systems similar to those advocated by groups like the Greywater Guerrillas are, however, so frequent in the South as to not be nearly as novel (Jimenez and Asano 2008). Here, in the South, water reuse is understood to be an absolute necessity not because it reduces the pollution impact of household water use, but because access is limited, literally making every drop count.

Because of the ease in which DIY water reuse in the Global North can be contextualized as inherently resistant to the established capitalist mandates of neoliberalism in the region, efforts to criminalize it are nearly inevitable; other forms of DIY reuse such as 'dumpster diving' for food and scavenging for discarded building materials have been criminalized in the developed world (Brisman 2009), and those who engage in such practices are increasingly socially marginalized as their lifestyles encroach on the demands made by consumer culture (Ferrell 2006). While the tendency of the state to work to protect the interests of capital by criminalizing many of the activities under the DIY, conservation, and reuse umbrella lays plain the state's perceived mandate to act as the protector of corporate interests, rather than a guardian of water purity, these DIY efforts illustrate that water activists are increasingly engaging in responses that conceptually merge issues of access *and* pollution rather than continuing to treat them along binary lines.

While the water justice movements of the Global South and those of the Global North are very different, each response is illustrative of the development of a truly global water consciousness concerned with the myriad ways in which the privatization of water, the removal of community-based controls on water supplies, and the criminalization of sustainable water practices and technologies jeopardize not only water access, but global water quality itself. Activists in the Global South, while currently concerned primarily with issues of inadequate access relating to privatization efforts, are increasingly likely to bear the brunt of water pollution, which is fuelled significantly by the problematic water habitus of the developed nations of the Global North (Shah 2006). In turn, water conscious activists in the Global North currently focus their attention on issues of pollution, although they are likely to find themselves facing many of the same issues of access currently prioritized in the Global South. Indeed, American activists concerned with water are already turning their attention to access issues. In *Graywater: The Next Wave*, primarily a design guide free of the explicitly political content that characterizes much water activism, Curtis McLamb briefly and prophetically expresses the access issues motivating his own implementation of greywater reuse; reflecting concerns over a coming global water shortage, McLamb (2004, p. 6) urges the use of greywater systems to deal with 'water shortages and restrictions … that can only get worse' Here, McLamb illustrates the willingness of the Global North to consider issues of access alongside issues of pollution and to tailor responses accordingly. The other side of the coin—the willingness of those active in the Global South to consider issues of pollution alongside issues of access—is illustrated by author and water activist Vandana Shiva, writing that 'more than anything, the oil economy's environmental externalities, such as pollution and climate change, will determine the future of water, and through the future of water, the future of all life' (Shiva 2002, p. 40). This recognition and exchange of concerns between the developed and developing worlds draw attention to the merging of issues of access and issues of pollution.

As climate change continues to advance, its impact on water issues—whether issues of pollution or access—will only increase. Because it is such a profoundly global phenomenon, the impact of climate change on water access and pollution provides a unique opportunity for various criminologies to consider water as a global, conceptual, and material site worthy of examination. As Agnew and others have noted, climate change is certain to have serious criminogenic effects relating to mass migration and displacement, increased strain, resource conflict and offender opportunity (Agnew 2012, pp. 15–17). Kramer (2014) has suggested that criminologists conceptualize climate change as state-corporate crime; adoption of that conceptual framework leads, almost inevitably, to the consideration of harmful water privatization schemes and regulatory repression of healthy water habitus as also criminogenic. With conflicts stemming from both issues—access and pollution—becoming increasingly likely, criminological and public imaginations would do well to consider these issues in tandem, rather than continuing to draw conceptual lines between the two.

Human Rights, Access, and Pollution: Concluding Thoughts on (E)merging Issues and Perspectives

In the decades following the initial proposals for a green criminology made by Lynch (1990), South (1998) and others, the field has integrated significant knowledge developed in the study of human rights. In particular, the environmental justice perspective within green criminology (for a discussion see White 2013) finds the roots of its thoughtful anthropocentrism in the environmental justice frameworks of social justice developed by Bullard (see generally 1983, 1994, 2005, 2005/2006; Bullard et al. 2001), which themselves are rooted in ideas flowing from human rights. Others working within the environmental justice perspective have continued to adopt human rights frameworks (see generally, Agyeman and Evans 2004; Boyle and Anderson 1996; Pellow 2007; Popovic 1996; Sachs 1996; Sachs and Peterson 1995). Because adequate and equitable access to clean water is essential to human life, the overlapping perspectives of human rights and environmental justice have frequently made water issues a key component of their shared discourse (see Bakker 2007; Farmer 2003; Gleick 1998; Pogge 2008). Human rights discourses relating to water as a necessary component of environmental justice may be concerned primarily with issues of pollution (Bullard 2012; Douglas et al. 2012) or issues of access (Bakker 2007; Gleick 1996, 1998). In the merging philosophical and discursive streams of environmental justice and human rights, we can again find reason to demolish the binary distinctions dividing issues of pollution from issues of access.

While access to clean water[4] may seem, on its surface, to be an unassailable human right—for without it, other rights cannot be realized—the wealthy hydromerchants of neoliberal global capitalism nevertheless rise to the challenge. In 2013, Peter Brabeck, the CEO of Nestle, the world's 27th largest corporation and a major privatizer of water, claimed in an interview that the idea that access to clean water was a human right was 'extreme'. While Brabeck later recanted—and Nestle employed a virtual army of public relations soldiers to attempt to turn back the clock on the CEO's candid statements—his second round of statements, meant to undo the public relations disaster caused by the initial interview, only doubled down on water privatization, as Brabeck insisted he only meant that water should always be assigned a market value to remind consumers of its natural value (Pantland 2013). The privatizing logic that animates Brabeck's perspective is certainly illustrative of a growing neoliberal tendency, but it can also be understood as the logical progression of what Marx (1867, p. 502) considered to be 'primitive accumulation', the processes by which populations and resources are co-opted by global capital.

Countering the logic of Brabeck and other corporate and institutional drivers of water privatization and pollution are perspectives that understand access to clean water as a fundamental human right. Indeed, the United Nations General Assembly codified these rights in 2010 with the adoption of resolution 64/92 which holds that there is a human right to 'safe and clean drinking water and sanitation' and that that access 'is essential for the full enjoyment of life and all human rights' (United Nations 2010). This resolution, and the activist responses that both contributed to its adoption by the UN and continue to underscore it, makes a plain fact of the interconnectedness of issues of water access and water pollution. While water activists in the North such as the Greywater Guerrillas and similar groups often make pollution the conceptual focus of their efforts, issues of pollution may have limited salience in regions where access to water is not assumed. Similarly, in the context of the Global North in which groups like the Greywater Guerrillas were founded, the salience of serious issues of access was limited by the artificial water abundance created by marketized provision systems. Here we can locate the origins of the conceptual division between issues of access and issues of pollution: the

[4] This simple summarizing phrase—*access to clean water*—captures the issue with binary conceptualizations of issues of access and pollution: access is only meaningful if it is access to unpolluted water, and pollution has the most recognized impact when it hinders access. While the latter may point to problematic issues of anthropocentrism (see generally Halsey and White 1998) and a tendency to ignore harms suffered by animals other than humans, it nonetheless holds true that humans will care the most about pollution events that hinder access to water that is materially or economically necessary to human health and development.

Northern world has long assumed that access is a given, seemingly making pollution a more relevant concern, while those in the South must first concern themselves with access and secondarily with pollution. In a more contemporary light, though, each issue may gain the globalized salience made possible by the human rights perspective. Because of the globalized and transnational nature of the human rights perspective, issues of access and pollution may be divorced from the spatial, social, and geopolitical distinctions that construct a problematic binary. By considering a globalized perspective informed by the idea of *access* to *clean* water as a human right, public and criminological imaginations can strive to recognize the global interconnectedness of water—an interconnectedness that transcends distinctions of state, place, and culture.

References

Abbey-Lambertz, K. (2014). How Detroit's water crisis is part of a much bigger problem. *The Huffington Post*, August 19, 2014. http://www.huffingtonpost.com/2014/08/19/detroit-water-shutoffs_n_5690980.html. Accessed 15 Feb 2015.

Agnew, R. (2012). Dire forecast: A theoretical model of the impact of climate change on crime. *Theoretical Criminology, 16*(1), 21–42.

Aguilar, M. (2014). 10 haunting photos from the Exxon Valdez oil spill catastrophe. *Reframe*, March 24, 2014. http://reframe.gizmodo.com/haunting-photos-from-the-exxon-valdez-oil-spill-catastr-1550691599. Accessed 15 Feb 2015.

Agyeman, J., & Evans, B. (2004). 'Just sustainability': The emerging discourse of environmental justice in Britain? *The Geographical Journal, 170*(2), 155–164.

Allen, L., & Woelfe-Erskine, C. (n.d.). About greywater action—for a sustainable water culture. http://greywateraction.org/content/about-us. Accessed 16 Mar 2013.

Associated Press. (2014a). Arizona town near Grand Canyon runs low on water. Associated Press/Yahoo! News. May 12. http://news.yahoo.com/arizona-town-near-grand-canyon-runs-low-water-062051573.html. Accessed 15 Feb 2015.

Associated Press. (2014b). Drought leads to water restrictions in Utah. *The Washington Times*, July 16, 2014. http://www.washingtontimes.com/news/2014/jul/16/drought-leads-to-water-restrictions-in-utah/. Accessed 15 Feb 2015.

Atkin, E. (2014). 25 years later, a heartbreaking look back at Exxon's Alaska oil spill. *Climate Progress*, March 24, 2014. http://thinkprogress.org/climate/2014/03/24/3418287/exxon-valdez-25-years-later/. Accessed 15 Feb 2015.

Bakker, K. (2007). The "commons" versus the "commodity": Alter- globalization, anti- privatization and the human right to water in the global south. *Antipode, 39*(3), 430–455.

Bardach, A. (2014). Lifestyles of the rich and parched. *Politico*, August 24, 2014. http://www.politico.com/magazine/story/2014/08/california-drought-lifestyles-of-the-rich-and-parched-110305.html. Accessed 15 Feb 2015.

Barlow, M. (2008a). *Blue covenant*. New York: The New Press.

Barlow, M. (2008b). Life, liberty, water. *Yes Magazine*, May 22, 2008. http://www.yesmagazine.org/issues/a-just-foreign-policy/life-liberty-water. Accessed 15 Feb 2015.

Barnett, C. (2014). Water works. *Utne Reader, 181*, 46–51.

Bernard, S. (2014). California drought leads to a black market for water. *Grist*, November 12, 2014. http://grist.org/news/california-drought-leads-to-a-black-market-for-water/. Accessed 15 Feb 2015.

Block, B. (2013). Arctic melting may lead to expanded oil drilling. *World Watch Institute*. http://www.worldwatch.org/node/5664. Accessed 15 Feb 2015.

Bostok, M., & Quealy, K. (2014). Mapping the spread of drought across the U.S. *The New York Times*. Updated July 31. http://www.nytimes.com/interactive/2014/upshot/mapping-the-spread-of-drought-across-the-us.html?_r=0. Accessed 15 Feb 2015.

Boyle, A., & Anderson, M. (Eds.) (1996). *Human rights approaches to environmental protection* (p. 313). Oxford: Clarendon Press.

Brisman, A. (2002). Considerations in establishing a stormwater utility. *Southern Illinois University Law Journal, 26*(3), 505–528.

Brisman, A. (2009). The indiscriminate criminalisation of environmentally beneficial activites. In R. White (Ed.), *Global environmental harm: Criminological perspectives* (pp. 161–192). Portland: Willan.

Brisman, A. (2013). Not a bedtime story: Climate change, neoliberalism, and the future of the arctic. *Michigan State International Law Review, 22*(1), 241–289.

Brisman, A., & South, N. (2013). A green-cultural criminology: An exploratory outline. *Crime Media Culture, 9*(2), 115–135.

Brisman, A., & South, N. (2014). *Green cultural criminology: Constructions of environmental harm, consumerism and resistance to ecocide*. London: Routledge.

Bullard, R. D. (1983). Solid waste sites and the black houston community. *Sociological Inquiry, 53*(2–3), 273–288.

Bullard, R. D. (1994). Environmental justice for all: It's the right thing to do. *University of Oregon Journal of Environmental Law and Litigation, 9,* 281–308.

Bullard, R. D. (Ed.) (2005). *The quest for environmental justice: Human rights, and the politics of pollution*. San Francisco: Sierra Club Books.

Bullard, R. D. (2005/2006). All transit is not created equal. *Race, Poverty, and the Environment, 12*(1), 9–12.

Bullard, R. D., Warren, R. C., & Johnson, G. S. (2001). The quest for environmental justice. In R. L. Braithwaite & S. E. Taylor (Eds.), *Health Issues in the black community* (2nd edn). San Francisco: Jossey-Bass Publishers.

Burke, G. (2014). Water flows uphill? Maybe, in California drought. Associated Press/Yahoo! News. May 7. http://news.yahoo.com/water-flows-uphill-maybe-california-drought-181712447.html. Accessed 15 Feb 2015.

Caplan-Bricker, N. (2013). This is what happens when a pipeline bursts in your town. *Newrepublic*, November 2013. http://www.newrepublic.com/article/115624/exxon-oil-spill-arkansas-2013-how-pipeline-burst-mayflower. Accessed 15 Feb 2015.

Cosier, S. (2014). A sickening swill. *On Earth*, December 11, 2014. http://www.onearth.org/earthwire/sickening-swill. Accessed 15 Feb 2015.

Crowley, K. (2014). South Africa gets funds for $ 900 million acid-water clean-up. *Bloomberg*, November 14, 2014. http://www.bloomberg.com/news/2014-11-14/south-africa-gets-funds-for-900-million-acid-water-clean-up.html. Accessed 15 Feb 2015.

Donn, J., Mendoza, M., & Pritchard, J. (2008). Pharmaceuticals lurking in US drinking water. *NBCNews.com*, March 10, 2008. http://www.nbcnews.com/id/23503485/ns/health-health_care/t/pharmaceuticals-lurking-us-drinking-water/. Accessed 15 Feb 2015.

Douglas, E. M., Kirshen, P. H., Paolisso, M., Watson, C., Wiggin, J., Enrici, A., & Ruth, M. (2012). Coastal flooding, climate change and environmental justice: Identifying obstacles and incentives for adaptation in two metropolitan Boston Massachusetts communities. *Mitigation and Adaptation Strategies for Global Change, 17*(5), 537–562.

Du Rees, H. (2001). Can criminal law protect the environment? *Journal of Scandinavian Studies in Criminology and Crime Prevention,* 2, 109–126.

Dwyer, L. (2014). Dry as a bone: Lake Mead's H2O situation just got a whole lot worse. *Takepart.com*. July 16. http://www.takepart.com/article/2014/07/16/lake-mead-water-drought-emergency. Accessed 15 Feb 2015.

Farmer, P. (2003). Pathologies of power: Health, human rights, and the new war on the poor. *North American Dialogue, 6*(1), 1–4.

Ferrell, J. (2006). *Empire of scrounge*. New York: New York University Press.

Food and Water Watch. (n.d.). Bottled water costs consumers and the environment. http://www.foodandwaterwatch.org/water/bottled/bottled-water-bad-for-people-and-the-environment/. Accessed 15 Feb 2015.

Gleick, P. H. (1996). Basic water requirements for human activities: Meeting basic needs. *Water International, 21*(2), 83–92.

Gleick, P. H. (1998). The human right to water. *Water Policy, 1*(5), 487–503.

Gonzalez Covarrubias, J. (2014). Mexico acid leak leaves orange river, toxic water. AFP/Yahoo! News. Aug. 16. http://news.yahoo.com/mexico-acid-leak-leaves-orange-river-toxic-water-050852546.html. Accessed 15 Feb 2015.

Greenpeace. (2013). Toxic threads: Polluting paradise. http://www.greenpeace.org/international/Global/international/publications/toxics/Water2013/Toxic-Threads-04.pdf. Accessed 15 Feb 2015

Halsey, M., & White, R. (1998). Crime, ecophilosophy, and environmental harm. *Theoretical Criminology, 2*(3), 345–371.

Hseih, J. (2014). Exxon Valdez memories from Alaska's capitol still fresh 25 years later. KTOO.org, March 24, 2014. http://www.ktoo.org/2014/03/24/memories-still-fresh-25-years-exxon-valdez/. Accessed 15 Feb 2015.

Jimenez, B., & Asano, T. (Eds.) (2008). *Water reuse: An international survey of current practice, issues and needs*. London: IWA Press.

Joling, D. (2014). 25 years later, Exxon Valdez spill effects linger. The Associated Press, March 20, 2014. http://www.cdispatch.com/news/article.asp?aid=31988. Accessed 15 Feb 2015.

Kramer, R. (2014). Climate change: A state-corporate crime perspective. In R. White, M. Kluin, & T. Spapens (Eds.), *Environmental crime and its victims: Perspectives within green criminology* (pp. 24–39). London: Ashgate.

La Franiere, S. (2006). In oil-rich Angola, cholera preys upon poorest. *The New York Times*, A1–14.

Los Angeles Times. (2014). Drought-plagued California gets tough on water wasters: Will it help? July 16, 2014. http://www.latimes.com/local/lanow/la-me-ln-drought-california-water-wasters-20140716-story.html. Accessed 15 Feb 2015.

Lynch, M. J. (1990). The greening of criminology: A perspective on the 1990s. *Critical Criminologist, 2*(3–4), 165–169.

Marx, K. (1867 [1976]). *Capital: A critique of political economy* (vol. 1) (trans: B. Fowkes). Harmondsworth: Penguin.

McClanahan, B. (2014). Green and grey: Water justice, criminalization and resistance. *Critical Criminology*. doi:10.1007/s10612-014-9241-8.

McClanahan, B., Brisman, A., & South, N. (In Press). Privatisation, pollution and power: A green criminological analysis of present and future water crises. In G. Barak (Ed.), *Routledge international handbook of crimes of the powerful*. London: Routledge.

McLamb, C. (2004). *Graywater: The next wave*. Suwanee: Graywater Resource.

Mok, O. (2014). Penang: Up to Putrajaya to punish IWK for water pollution. MalayMail Online. March 3, 2014. http://www.themalaymailonline.com/malaysia/article/penang-up-to-putrajaya-to-punish-iwk-for-water-pollution. Accessed 15 Feb 2015.

Morin, M. (2013). Ice melt to expand Arctic shipping by midcentury. *Los Angeles Times*, March 04, 2013. http://articles.latimes.com/2013/mar/04/science/la-sci-sn-arctic-shipping-lanes-20130304. Accessed 15 Feb 2015.

Murphy, S. (2014). Coca-Cola and Nestle are sucking us dry without our even knowing. The Motley Fool. Mar. 2. http://www.fool.com/investing/general/2014/03/02/coca-cola-and-nestle-are-sucking-us-dry-without-us.aspx. Accessed 15 Feb 2015.

Nurse, A. (2013). Cleaning up greenwash: A critical evaluation of the activities of oil companies in The Niger Delta. https://www.northumbria.ac.uk/static/5007/sasspdf/Nurse.pdf. Accessed 15 Feb 2015.

Omotola, S. (2009). Dissent and state excesses in the Niger Delta, Nigeria. *Studies in Conflict & Terrorism, 32*(20), 129–147.

One World South Asia. (2006). Indian water activists launch anti-privatization campaign. February 9, 2006. http://infochangeindia.org/water-resources/news/indian-water-activists-launch-anti-privatisation-campaign.html. Accessed 15 Feb 2015.

Osnos, E. (2014). Chemical valley. *The New Yorker,* Apr. 7, 38–49.
Pantland, W. (2013). Nestle chairman Peter Brabeck makes the astonishing claim that water isn't a human right. USI Live, October 1, 2013. https://usilive.org/nestle-ceo-says-water-isnt-a-human-right-tell-him-hes-wrong/. Accessed 15 Feb 2015.
Paulson, N., Zagorski, K., & Ferguson, D. C. (In Press). On harm and mediated space: The BP oil spill in the age of globalization. In A. Brisman, N. South, & R. White (Eds.), *Environmental crime and social conflict: Contemporary and emerging issues*. Surrey: Ashgate.
Pellow, D. (2007). *Resisting global toxics: Transnational movements for environmental justice*. Boston: MIT Press.
Piper, K. (2006). *Left in the dust: How race and politics created a human and environmental tragedy in L.A.* New York: Macmillan.
Pogge, T. W. (2008). *World poverty and human rights*. London: Polity.
Popovic, N. (1996). Pursuing environmental justice with international human rights and state constitutions. *Stanford Environmental Law Journal, 15,* 338–379.
Powell, B. (2014). Water: The price is wrong. *The Huffington Post*. June 20. http://www.huffingtonpost.com/ben-powell/water-the-price-is-wrong_b_5512504.html. Accessed 15 Feb 2015.
Prüss-Üstün, A., Bos, R., Gore, F., & Bartram, J. (2008). Safer water, better health: Costs, benefits and sustainability of interventions to protect and promote health. Report prepared for World Health Organization. http://whqlibdoc.who.int/publications/2008/9789241596435eng.pdf. Accessed 15 Feb 2015.
Reid, T. (2014). In California drought, big money, many actors, little oversight. Reuters/Yahoo! News. Apr. 25. http://news.yahoo.com/california-drought-big-money-many-actors-little-oversight-110415148–finance.html. Accessed 15 Feb 2015.
Reisner, M. (1993). *Cadillac desert* (revised ed.). New York: Penguin.
Sachs, A. (1996). Upholding human rights and environmental justice. In L. R. Brown (Ed.), *State of the world 1996: A worldwatch institute report on progress toward a sustainable society* (pp. 133–151). New York: W. W. Norton.
Sachs, A., & Peterson, J. A. (1995). *Eco-justice: Linking human rights and the environment* (pp. 1–68). Washington, DC: Worldwatch Institute. (Paper 127).
Shah, A. (2006). Water and development. GlobalIssues.org. http://www.globalissues.org/article/601/water-and-development. Accessed 15 Feb 2015.
Shiva, V. (2002). *Water wars*. Cambridge: South End Press.
Smith, S. (2014). California drought spawns well drilling boom. Associated Press/Yahoo! News. Apr. 13. http://news.yahoo.com/california-drought-spawns-well-drilling-boom-143021112.html. Accessed 15 Feb 2015.
South, N. (1998). A green field for criminology? A proposal for a perspective. *Theoretical Criminology, 2*(2), 211–233.
South, N., & Walters, R. (2014). Eco-crimes and fresh water. Paper presented at the British Society of Criminology Annual Conference 2014. University of Liverpool, Liverpool, UK (July 10, 2014).
South, N., Brisman, A., & McClanahan, B. (2014). Green criminology. *Oxford bibliographies online: Criminology*. doi:10.1093/OBO/9780195396607-0161.
Steinmentz, K. (2014). Unprecedented california drought restrictions go into Effect. TimeMagazine.com, July 29, 2014. http://time.com/3048600/california-drought-fines-water-rules/. Accessed 15 Feb 2015.
Strauss, G. (1999). Ten years later, case is hardly closed. USA Today, March 4, 1999. Money section: 1. http://www.jomiller.com/exxonvaldez/usatoday.html. Accessed 15 Feb 2015.
Stretesky P. B., Long, M. A., & Lynch, M. J. (2013). Does environmental enforcement slow the treadmill of production? The relationship between large monetary penalties, ecological disorganization and toxic releases within offending corporations. *Journal of Crime and Justice, 36*(2), 233–247.
Taylor, A. (2014). The Exxon Valdez oil spill: 25 years ago today. *The Atlantic*, March 24, 2014. http://www.theatlantic.com/infocus/2014/03/the-exxon-valdez-oil-spill-25-years-ago-today/100703/. Accessed 15 Feb 2015.

Torres-Leclercq, B. (2014). Greenpeace Chile plants flag in Andes to create new 'Glacier Republic'. *The Santiago Times*. March 06, 2014. http://santiagotimes.cl/greenpeace-chile-plants-flag-andes-create-new-glacier-republic/. Accessed 15 Feb 2015.

U.S. Drought Monitor. (2014). Michigan data. http://droughtmonitor.unl.edu/Home/StateDroughtMonitor.aspx?MI. Accessed 15 Feb 2015.

United Nations. (2010). General assembly Resolution 64/292: The human right to water and sanitation. http://www.dgvn.de/fileadmin/user_upload/DOKUMENTE/English_Documents/A-Res-64-292.pdf. Accessed 15 Feb 2015.

Waldie, D. J. (2014). Drought by the numbers: Where does California water go? KCET, February 10, 2014. http://www.kcet.org/updaily/socal_focus/commentary/where-we-are/in-a-season-of-drought-where-does-the-water-go.html. Accessed 15 Feb 2015.

Walker, A. (2014). Rich people are trucking their own water into drought-ridden california. Gizmodo.com, August 26, 2014. http://gizmodo.com/rich-people-are-trucking-their-own-water-into-drought-r-1627178060. Accessed 15 Feb 2015.

White, R. (2005). Environmental crime in global context: Exploring the theoretical and empirical complexities. *Current issues in criminal justice, 16*(3), 271–285.

White, R. (2013). *Environmental harm: An eco-justice perspective*. Bristol: Policy Press.

White, R., & Heckenberg, D. (2014). *Green criminology: An introduction to the study of environmental harm*. London: Routlege.

Woody, T. (2014a). California's catastrophic drought just got worse—a lot worse. Takepart.com/Yahoo! News. June 20. http://news.yahoo.com/californias-catastrophic-drought-just-got-worse-lot-worse-181115468.html. Accessed 15 Feb 2015.

Woody, T. (2014b). The drought apocalypse approaches as the Colorado river basic dries up. Takepart.com/Yahoo! News. July 24. http://news.yahoo.com/drought-apocalypse-approaches-colorado-river-basin-dries-221214264.html. Accessed 15 Feb 2015.

Bill McClanahan is a PhD candidate at the University of Essex in Colchester, UK.

Chapter 6
Exploring Environmental Inequality Within US Communities Containing Coal and Nuclear Power Plants

Sarah Kosmicki and Michael A. Long

Introduction

Employing one of the world's most advanced electricity generating and distributing systems, the USA consumed approximately 3856 billion kWh of electricity in 2011 (USEIA 2012a). While there are clear benefits to our nation's widespread energy accessibility, this resource comes at a cost to public and environmental health. Over 60 % of our national electricity is produced from coal and nuclear power which often employ environmentally hazardous methods of sourcing, extracting, and generating electricity (USIEA 2012a). Coal-fired power plants require massive amounts of coal, most commonly extracted on USA soil through surface mining techniques including mountain top removal (USEIA 2012b). This mining method requires blasting open mountains with explosive material to uncover coal beneath the earth's surface. The repercussions of mountaintop removal include the ecological devastation of entire habitats through the leveling of landscapes and the poisoning of ecosystems and human communities from chemical runoff (Epstein and Reinhart 2010). When coal is burned for the production of electricity, greenhouse gases are released into the atmosphere where they contribute to changes in global climate patterns. Further, pollutants such as particulates and noxious gasses released during combustion from coal-fired energy production have grave implications for public health. For instance, exposure to particulates has been associated with heart attacks, strokes, and premature death (Keating 2004). Additionally, gasses such as nitrous oxides and sulfur dioxide contribute to cardiovascular and respiratory diseases, and mortality (Samoli et al. 2006; Sunyer et al. 2003).

M. A. Long (✉)
Department of Social Sciences & Languages, Northumbria University, Lipman Building, Newcastle upon Tyne NE1 8ST, UK
e-mail: michael.long@northumbria.ac.uk

S. Kosmicki
Oklahoma State University, Stillwater, USA

T. Wyatt (ed.), *Hazardous Waste and Pollution,* DOI 10.1007/978-3-319-18081-6_6

Nuclear power production poses potentially greater risks than those of coal-fired energy. The process begins with the mining of uranium, which is harmful to human health and has been associated with adverse effects of the cardiovascular and nervous systems (Taylor and Taylor 1997). Research has also found that communities living in close proximity to a nuclear powered facility have higher rates of infant mortality and are at an increased risk of suffering from breast cancer and leukemia (Gould 1996; Mangano 2002; Swings et al. 1989). Further, as revealed by Three Mile Island, Chernobyl, and more recently, Fukushima, the production of nuclear power has the potential for devastating impacts in the form of nuclear meltdowns. Last, nuclear waste is a continually growing hazard for which there are currently no viable solutions for safe disposal.

As a result of the dangers posed by our nation's energy cycle, researchers (Brulle and Pellow 2006; Bryant and Mohai 1992; Bullard 1990; Faber and Krieg 2002; Grineski et al. 2010), advocates (Energy Action Coalition 2012; Energy Justice Network 2012; Indigenous Environmental Network 2012), and locally impacted citizens are seeking energy justice. Like other environmentally hazardous activities, the mining of coal and uranium and the disposal of coal ash and nuclear waste have attracted the attention of many environmental justice scholars. Many of these researchers have determined that poor, minority, and rural populations, including Native American communities, are disproportionately impacted by uranium mining and nuclear waste storage sites (Brugge and Goble 2002; Gerrard 1994; Malin and Petrzelka 2010; Markstrom and Charley 2003; Taliman 1992). Similarly, findings suggest that poor and rural communities are more likely to live in the USA coal fields and suffer negative impacts associated with mountaintop mining (Evans 2010; Hendryx 2011; McGinley 2004). In contrast to research on mining and disposal, environmental justice researchers have largely ignored inequalities concerning electricity generating power plants. Given the public health impacts associated with exposure to coal and nuclear powered stations, this gap in the literature is rather surprising.

While limited research has found that poor and minority communities are unequally impacted by coal-fired power plants, it is not clear whether these populations are disproportionately burdened by nuclear power plants (Faber and Krieg 2002; Keating and Davis 2002). A study seeking to identify populations most likely to host nuclear power plants found no environmental inequality at the national level (Alldred and Shrader-Frechette 2009). This finding may suggest that variables beyond income and race help account for disproportionate exposure to environmental hazards. It may also indicate that nuclear power plants are not conceptualized as dangerous sites in a way that other hazardous facilities are. Accordingly, this chapter seeks to answer the following questions: (1) Are minorities more likely to live near coal-fired or nuclear power plants?; (2) Are poverty and income indicators of living in close proximity to coal-fired or nuclear power plants; and (3) Are children more likely to live near a coal-fired or nuclear power plant?

The remainder of the chapter is organized as follows. First, we review literature on risk society, the treadmill of crime and environmental justice focusing on links with coal-fired and nuclear power plants. Next, we provide our hypotheses, followed by our methodology and results. The discussion and conclusion situates our findings in the existing literature and highlights directions for future research.

Background

Risk Society and Risk Perception

Although, the production of electricity from coal and nuclear power has serious impacts on human and ecosystem health, these costs are generally deemed acceptable in light of the energy-intensive lifestyle of our nation. Risk society theory informs us that with increasing modernization, there follows an increase in the number and severity of accompanying risks. From the widespread use of hazardous chemicals to the changes in global weather patterns, risks from our increasingly industrialized society are growing exponentially. Ulrich Beck (1992) coined the phrase "risk society" to refer to a society that is economically dependent upon intensive modernization, and consequently, vulnerable to possible negative outcomes. Thus, given the consumption patterns of our nation, risks posed from energy production tend to be viewed as necessary consequences of our modern lifestyles. In order to maintain a particular standard of living, energy demands must be satisfied and the consequent risks are outweighed by the advantages of modernization. Therefore, it is within this context that we can understand the assent of harms associated with the production of US energy.

Risk society theory is further useful in understanding the impacts of energy production as power plants pose risks that are not entirely understood and have the potential to be catastrophic. While in the past it was believed that the consequences of modernization could be controlled and compensation could be provided to those harmed by negative externalities, it has become evident that the hyper modernization of our present day has led to the formation of risks which cannot be so easily managed (Beck 2006). These risks are distinct from those in earlier societies in that they are delocalized temporally, spatially, and socially, and are incalculable and non-compensable (Beck 2006). Coal-fired and nuclear power plants provide striking examples of the sources of risk arising from modernity.

Coal-fired stations pose spatially delocalized risks as they are major emitters of greenhouse gas emissions and thus contribute to global climate change. Changes in global climate patterns may be irreversible, thus impacts of coal plants are temporally delocalized. Further, coal plant companies are only one of the many greenhouse gas contributors, and thus risks from coal plants are socially delocalized as companies will not be held socially or financially responsible for the consequences of emissions. Next, risks from coal plant releases are incalculable since shifting weather patterns are not fully understood and cannot be reliably quantified. Last, the impacts of these emissions may be non-compensable as changes in the global climate may create a planet no longer hospitable to human life.

In the case of nuclear power plants, the impact of a nuclear meltdown would transcend geographic and political boundaries. Additionally, certain radionuclides would persist in affected areas over several generations. For example, some of the radionuclides released during a nuclear accident would decay over a period of several days or weeks, while others such as plutonium-239 and plutonium-240 have half-lives of thousands of years (Nuclear Energy Agency 2002). First, this risk is delocalized socially as the ambiguity of impacts from a nuclear accident makes it difficult to identify and hold liable the responsible parties. Second, the potential impact of an accident cannot be calculated, and thus it cannot be known whether the potential consequences outweigh the benefits of nuclear power. For instance, scientists are unable to determine how many people would suffer specific adverse health outcomes as a result of a nuclear meltdown. Indeed, this feat is impossible even after an accident has occurred as there is no way to definitely determine that a specific individual's compromised health was a direct result of exposure to radiation. Third, a nuclear meltdown is non-compensable as harmed individuals cannot be compensated for loss of life or health.

The Treadmill of Crime

As a result of the hyper modernization characteristic of risk society, there arose previously unforeseen consequences to the environment. This modernization is also a critical component of the treadmill of production theory which has been used by green criminology scholars (Long et al. 2012; Stretesky et al. 2013) to situate the role of political economy in crimes against nature.

The treadmill of production theory was crafted by Schnaiberg (1980) to describe how political economy impacts the destruction of the natural environment. The treadmill is propelled by five primary objectives (Schnaiberg et al. 2002). First, our capitalist society demands constant economic expansion, and thus, producers must maximize efficiency and increase their profits in order to survive in the system. With no ceiling for economic growth, the primary objective of a capitalist society is to increase profits. Second, much of this growth is expected to result from large firms who have the capital to invest in further growth and supply jobs to consumers, who then have the means to increase their personal consumption. Third, as economic growth occurs from the increased production of goods, the system requires an increase in consumption to perpetuate the cycle. Fourth, this cycle is kept in balance from alliances between capital, workers, and the government. Fifth, the system seeks to resolve environmental and social dilemmas through further economic expansion. For example, speeding the treadmill can reduce unemployment by increasing the number of available jobs while environmental damages are expected to be repaired by the development of new technologies (Schnaiberg et al. 2002). These goals cannot be achieved without increasing the amount of energy produced, and thus power plants play a vital role in the sustenance and perpetuation of the treadmill and, consequently, in harms against nature.

Coal and nuclear power plants are an important component of the treadmill of production not only because the product they generate fuels the system but also their operations are associated with extensive ecological withdrawals and additions. The primary withdrawal associated with coal-fired power plants is the mining of coal. While the USEIA (2012c) projects that the USA possesses enough coal to meet the nation's energy needs for 200 years, the treadmill assumes increasing rates of production, and therefore, the finite amount of coal within the earth's surface will be unable to sustain the system for this period of time. Additionally, resource depletion is not the only consequence of coal extraction. The mining of coal is often done through surface mining techniques including mountain top removal that results in the leveling of mountains, extensive deforestation, and the contamination of local streams (Long et al. 2012). Similarly, the mining of uranium required for the production of nuclear energy requires massive withdrawals from the environment. As uranium exists in small concentrations within the earth's surface, collecting a single ton of uranium requires the mining and processing of 98,000 tons of rock (Hughes 2006). Additionally, uranium is a finite resource which is expected to be exhausted in 20 years should worldwide nuclear power production double as some have predicted (Hughes 2006).

This chapter, however, is more concerned with the additions associated with coal and nuclear power plant operations because populations living near plant sites are more likely to be impacted by additions than by withdrawals. For instance, when coal is burned at a power plant, undesirable byproducts are produced along with electricity. These additions include particulate matter, noxious gasses, and mercury which have been associated with adverse health outcomes including heart attacks, strokes, neurological disorders, and premature death (Keating 2004; Samoli et al. 2006; USEPA 2012b). Similarly, the production of nuclear power requires the addition of hazardous substances, primarily in the form of radioactive waste. Most of this waste is stored in large steel and concrete pools at nuclear plant sites where it is expected to be safely stored for 120 years (Nuclear Energy Institute 2012). However, leaks occur during routine plant operations. Small amounts of radiation have been associated with increased incidences of cancer and infant mortality (Clapp et al. 1987; Gould 1996; Mangano 2008).

While the benefits of energy accessibility are shared by most citizens, the harmful additions associated with electricity production are not so equally distributed. Like many activities in the treadmill of production, the generation of electricity through coal-fired and nuclear power plants has greater impacts on select populations. In the case of power plant sites, communities living in close proximity suffer an unequal burden of adverse health outcomes from pollutant exposure.

Environmental Justice

In 2011 coal-fired power plants produced approximately 42% of the electricity consumed in the USA (USEIA 2012a). Not only is coal the least expensive source of electricity production in the nation but it is also abundant in the USA (USEIA 2012a). However, generating electricity by burning coal has severe environmental and public health ramifications. Coal-fired plants produce approximately 27 % of carbon dioxide (CO_2) emissions in the nation, making it a major contributor to greenhouse gas emissions (USEPA 2012a).

Coal-fired power plants produce large quantities of particulate matter. Exposure to particulate matter, especially finer particulates, has been shown to cause inflammation of the cardiovascular and respiratory systems (Keating 2004; USEPA 2012b). Particulate matter exposure has also been associated with premature and low birth weight infants and sudden infant death syndrome, making pregnant women and children particularly vulnerable to adverse health outcomes (Keating 2004). Coal-fired power plants further impact the health of women and children living in close proximity as they are responsible for producing more than 40 % of our nation's mercury emissions. Exposure to mercury has been associated with disruptions in neurological development in fetuses, babies, and children; and irregular pulse, blood pressure, neurologic, and kidney disorders in adults (Keating 2004; USEPA 2012b).

Hazardous gas emissions from coal powered plants also possess serious implications for public health. Research has found a strong association between exposure to NOx and cardiovascular, and respiratory mortality (Samoli et al. 2006). Children exposed to the gas may experience adverse health effects such as colds, coughs, and sore throats (Pilotto et al. 1997). When released into the atmosphere nitrous dioxide (NO_2) may react with hydrocarbons and sunlight to form a secondary pollutant, tropospheric ozone (Brunekreef and Holgate 2002). Exposure to ozone has been associated with diminished lung function, respiratory illnesses, and cardiovascular and respiratory deaths (Galizia and Kinney 1999; Gryparis et al. 2004). Similarly, exposure to another power plant gas, SO_2, has been positively correlated with cardiovascular disease related hospital admissions (Ibald- Mulli et al. 2001; Sunyer et al. 2003). Additionally, the release of SO_2 results in the formation of acidic particulates which have been linked to an increased incidence of bronchitis in children (Dockery et al. 1996).

Given the dangers associated with coal-fired power plant pollutants, it is not surprising that these facilities have been called the most harmful industrial air polluters in terms of impacts on human health and the natural environment (Keating and Davis 2002; Schneider and Banks 2010). Emissions from coal-fired power plants alone are estimated to have resulted in 9700 hospital admissions, 12,300 emergency room visits for asthma, 20,400 nonfatal heart attacks, and 13,200 mortalities in the year 2010. The monetary cost of these health outcomes is estimated at $ 100 billion per year (Schneider and Banks 2010). The brunt of this health and economic burden is shouldered by communities living in close proximity to coal plants.

While nuclear power plants serve a similar function to coal-fired plants, they have several important distinctions. Nuclear plants provide approximately 20 % of the USA electricity. However, this energy is generated at only 64 sites in the nation, compared to the 560 coal-fired plants in the USA (USEIA 2012b). Additionally, hazards posed by nuclear facilities are distinct from risks posed by coal-fired plants. Theoretically, nuclear power generation should not produce any emissions, although there are some releases of carbon dioxide that result from activities involving the maintenance of the facility. Rather, the public health dangers posed by nuclear power plants involve errors in some part of the production and storage of nuclear waste. These accidental releases are rarely catastrophic and generally involve small amounts of escaped radiation during routine procedures or leaks in storage containers (USEIA 2012b). While these releases may be minute, this does not imply that they are harmless to human health. A report by the National Academy of Science (2005) determined that there is no safe threshold for exposure to radiation, and even very small amounts of exposure have the potential to result in cancer. This finding has grave implications for those living near nuclear power facilities as various studies have found increased rates of breast cancer, leukemia, childhood cancers, and infant mortality in communities surrounding nuclear reactors (Clapp et al. 1987; Gould 1996; Mangano 2008; National Academy of Sciences 2005; Swings et al. 1989).

Over the past three decades, the growing body of environmental justice research has increasingly found that poor and minority communities in the USA are disproportionately impacted by the risks posed from hazardous environmental sites. For instance, studies have revealed that these populations are unequally located near hazardous waste facilities, incinerators, manufacturing facilities, and landfills (Bullard 1990; Downey 2005; Faber and Krieg 2002; Mohai and Bryant 1992; Pellow 2000). However, studies have yet to evaluate disproportionate exposure to nuclear power plants despite the fact that hazards associated with living near a nuclear plant are well documented. One exception is the work of Alldred and Shrader-Frechette (2009) who examined whether zip codes containing nuclear plants had higher percentages of poor and minority populations than non-host zip codes with results indicating no environmental injustice at the national level. Otherwise, the majority of studies examining communities living in close proximity to nuclear plants focus on risk perceptions (Eiser et al. 1995; Hung and Wang 2011; Sjoberg 2003; Stone 2001; Venables et al. 2009).

Conversely, previous research, although limited, indicates that coal-fired plants are unequally located in lower socioeconomic and minority communities (Faber and Krieg 2002). Indeed, one study found that 68 % of African Americans in the USA live within 30 miles of a coal-fired power plant compared to only 56 % of whites (Keating and Davis 2002). Further, research suggests that poor and minority populations are more likely to live near heavier polluting facilities compared to whiter and wealthier populations (Faber and Krieg 2002). Additionally, certain minority groups are at a greater risk of exposure due to lifestyle activities.

In addition to poor and minority populations, children are also particularly susceptible to coal plant emissions. For example, research has indicated that hazardous facilities tend to be located in communities with populations containing high percentages of children (Grineski et al. 2010; Perlin et al. 2001). Children are further vulnerable as their lungs are still developing and are larger in volume relative to body size than adults. Additionally, their immune systems are not fully developed and thus their bodies may react differently to pollutants compared to a healthy adult.

Further, children have higher rates of respiration than adults and are more likely to spend time outside engaging in physical activity, both of which increase the inhalation of air pollutants (Hill and Keating 2002).

Based on the preceding discussion, we formulate the following eight hypotheses:

Hypothesis 1 Census tracts hosting coal-fired power plants contain greater percentages of minorities compared to non-host census tracts.

Hypothesis 2 Census tracts hosting coal-fired power plants contain lower median household incomes compared to non-host census tracts.

Hypothesis 3 Census tracts hosting coal-fired power plants contain greater percentages of families living below the poverty line compared to non-host census tracts.

Hypothesis 4 Census tracts hosting coal-fired plants contain higher percentages of children than non-host census tracts.

Hypothesis 5 Census tracts hosting nuclear power plants do not contain greater percentages of minorities compared to non-host census tracts.

Hypothesis 6 Census tracts hosting nuclear power plants do not contain lower median household incomes compared to non-host census tracts.

Hypothesis 7 Census tracts hosting nuclear power plants do not contain greater percentages of families living below the poverty line compared to non-host census tracts.

Hypothesis 8 Census tracts hosting nuclear plants contain higher percentages of children than non-host census tracts.

Methodology

Data

The dependent variable is *Tract Type* with categories *non-host tract, coal host tract, nuclear host tract*, and *both host tract*. Host tracts were operationalized as census tracts with 50% or more of their area falling within a selected buffer around plant sites. All other tracts were categorized as non-host tracts.

The data for the independent variables was derived from the 2010 census provided by American Fact Finder and included *percent of the population that is minority, median household income, percent of families living below the poverty line,* and *percent of the population that are children.* To evaluate the relationship between race and proximity to power plants, percent of the population that is minority was calculated for every census tract in the nation. For the purpose of this study, any race other than white was categorized as minority. In order to evaluate relationships between class and proximity to coal and nuclear power plants, this study examined median household income and percentage of families living below the poverty line for each tract. Last, as previous studies have indicated that children are disproportionately impacted by environmental hazards, we calculated the percent of the population that is children aged 15 years and younger for each census tract (Grineski et al. 2010; Perlin et al. 2001).

This study also necessitated additional control variables. To begin with, urban areas are associated with both higher percentages of people of color and the presence of environmentally hazardous facilities. Thus, *population density,* reported as people per square mile, was included as a control variable. Additionally, certain socioeconomic variables have been indicated as significant predictors of proximity to environmentally hazardous facilities and were accounted for in this study. For instance, Anderton et al. (1994) found that employment in the manufacturing industry was more closely associated with living near hazardous waste treatment, storage, and disposal facilities than race and income. Thus, we controlled for relationships between occupation and proximity to plant by including the variable *percent of the population employed in the utilities sector* to account for employees of coal and nuclear powered plants. This variable represents the percent of the employed civilian population 16 years and older that have occupations in the transportation, warehouse, and utility sectors. Additionally, level of education has been associated with unequal exposure to environmental harms with less educated communities at greater risk (Mohai and Saha 2006). Bullard (1990) postulates that populations with greater resources including education are better able to successfully fight the placement of hazardous facilities in their communities. Therefore, to control for level of education we examined the *percent of the population with a bachelor's degree or above* for all census tracts. Last, as Alldred and Shrader-Frechette (2009) found disparities in demographics surrounding nuclear plants in the south, *region* was included as a variable in this study. Tracts were categorized as being in the West, Midwest, Northeast, or South according to the regions defined by the US census.

Method

To spatially analyze the data, we used a map of all census tracts in the USA containing demographic census data (United States Census Bureau 2010 Census TIGER/Line Shapefiles 2012). Coordinate information for all coal-fired and nuclear power plants in the nation was accessed through the eGRID database and plotted on the base map. Using ArcMap 10 buffer tool, buffer zones were drawn around each of the plants. We employ a 10 mile buffer zone in order to focus on populations most affected by pollutants.

After drawing buffers around plant sites, we used the ArcMap Clip function to extract the census data from all tracts with any of their area included in a buffer. We determined qualifying tracts by comparing the size of the original tract to the size of the clipped tract fragment. Census tracts with half of their area falling within a buffer zone were labeled as coal host, nuclear host, or both host tracts if the buffer was surrounding a coal plant, nuclear plant, or both a coal and nuclear plant respectively. All other tracts were labeled as non-host tracts. The data extracted from this map contained the data for all of the variables.

Analysis and Results

Table 6.1 contains the descriptive statistics for all variables in the analysis. Tables 6.2 and 6.3 provide the results of multinomial logistic regression models predicting tract type. Model 1 estimates the effects of demographic variables on the probability

Table 6.1 Descriptive statistics for independent and dependent variables

Independent variables	Mean	SD	Min	Max	Description
Minorities	36.07	30.12	0	100	Percent population minorities
Median income	55,506.30	26,885.80	5000	249,194	Median household income
Poverty	11.38	11.15	0	100	Percent families living below poverty line
Bachelors	27.00	18.22	0	100	Percent population with bachelor's degree
Children	20.72	5.70	0	57.62	Percent population children
Population density	5254.96	11,747.80	0.03	508,698	People per square mile
Utilities	5.12	3.45	0	63.20	Percent population employed in the utilities sector
Region					
West	0.22	0.42	0	1	West=1, else=0
Midwest	0.23	0.42	0	1	Midwest=1, else=0
South	0.36	0.48	0	1	South=1, else=0
Northeast	0.19	0.39	0	1	Northeast=1, else=0
Dependent variable	Frequency	Percent			
Tract type					
Non-host tract	56,662	78.62			
Coal host tract	14,514	20.14			
Nuclear host tract	571	0.79			
Both host tract	321	0.45			
Total	72,068	100			

N=72,068, Data comes from the 2010 US Census

Table 6.2 Multinomial logistic regression analysis examining the effect of demographic variables on the probability of living in a tract located within 10 miles of a coal plant, nuclear plant, or both

	Model 1[a]					
	Coal host		Nuclear host		Both host	
	B	*SE*	*B*	*SE*	*B*	*SE*
Minority	0.00762***	0.00044	−0.00108	0.0024	−0.00870*	0.0035
	(1.008)		(0.999)		(0.991)	
Median Income[b]	−0.00802***	0.00073	0.00121	0.0029	−0.00668	0.0041
	(0.992)		(1.001)		(0.993)	
Poverty	0.0105***	0.0013	−0.0205**	0.0075	−0.045***	0.011
	(1.011)		(0.980)		(0.956)	
Bachelors	0.0181***	0.00092	0.00140	0.0043	0.00919	0.0059
	(1.018)		(1.001)		(1.009)	
Children	−0.0123***	0.0022	0.0122	0.011	0.0765***	0.015
	(0.988)		(1.012)		(1.080)	
Population density	0.0148***	0.00090	−0.130***	0.020	−0.0916***	0.024
	(1.015)		(0.878)		(0.912)	
Utilities	0.0377***	0.0030	0.0879***	0.010	0.113***	0.014
	(1.038)		(1.092)		(1.119)	
West[c]	−0.920***	0.039	−1.839***	0.22	−16.161	417.86
	(0.399)		(0.159)		(9.58e-8)	
Midwest[c]	1.494***	0.026	−0.354***	0.13	0.394*	0.16
	(4.453)		(0.702)		(1.483)	
Northeast[c]	0.873***	0.030	0.859***	0.11	1.334***	0.15
	(2.394)		(2.361)		(3.795)	
Constant	−2.368***	0.057	−4.824***	0.24	−6.650***	0.34
LR Chi²(30)	9979.07***					
Cragg-Uhler R²	0.189					
AIC	72860.042					
BIC	73163.159					

[a] Coefficients versus the base category of non-host. Relative risk ratios are in parentheses
[b] Median income and population density are reported in thousands
[c] Effect of region with South as base category
**p*<0.05, ***p*<0.01, ****p*<0.001.

of tracts being located within a 10 mile radius of a plant compared to being located outside the buffer. As predicted, this model indicates that communities which contain higher percent population minority are more likely to be located in a coal host tract versus a non-host tract (Relative Risk Ratio = 1.008; $p < 0.001$). Additionally, populations living within a coal host tract are more likely to have lower median household incomes (RRR = 0.992; $p < 0.001$) and higher percent population families living below the poverty line (RRR = 1.011; $p < 0.001$) than non-host tracts. Tracts

Table 6.3 Multinomial logistic regression analysis examining the effect of demographic variables on the probability of living in a tract located within 10 miles of a nuclear plant or both a coal and nuclear plant

	Model 2[a]				Model 3[b]	
	Nuclear host		Both host		Both host	
	B	*SE*	*β*	*SE*	*β*	*SE*
Minority	−0.00870***	0.0024	−0.0163***	0.0036	−0.00761	0.0043
	(0.991)		(0.984)		(0.992)	
Median income[c]	0.00922**	0.0030	0.00134	0.0041	−0.00789	0.0050
	(1.009)		(1.001)		(0.992)	
Poverty	−0.0310***	0.0075	−0.0558***	0.012	−0.0248	0.014
	(0.969)		(0.946)		(0.976)	
Bachelors	−0.0167***	0.0044	−0.00889	0.0059	0.00780	0.0072
	(0.983)		(0.991)		(1.008)	
Children	0.0245*	0.011	0.0889***	0.015	0.0643***	0.018
	(1.025)		(1.093)		(1.066)	
Population density	−0.145***	0.020	−0.106***	0.024	0.0388	0.031
	(0.865)		(0.899)		(1.040)	
Utilities	0.0502***	0.011	0.0751***	0.015	0.0249	0.017
	(1.051)		(1.078)		(1.025)	
West[d]	−0.919***	0.23	−15.241	417.86	−14.322	417.86
	(0.399)		(2.40e-7)		(6.03e-7)	
Midwest[d]	−1.847***	0.13	−1.100***	0.16	0.748***	0.21
	(0.158)		(0.333)		(2.112)	
Northeast[d]	−0.0140	0.11	0.461**	0.15	0.475**	0.18
	(0.986)		(1.585)		(1.607)	
Constant	−2.455***	0.25	−4.282***	0.35	−1.826***	0.42

[a] Coefficients versus the base category of coal host
[b] Coefficients versus the base category of nuclear host. Relative risk ratios are in parentheses
[c] Median income and population density are reported in thousands
[d] Effect of region with South as base category
$*p<0.05$, $**p<0.01$, $***p<0.001$

with greater percent population with a bachelor's degree or above tend to be coal host versus non-host tracts (RRR = 1.018; $p<0.001$). Coal host tracts are also associated with higher population density (RRR = 1.015; $p<0.001$) and higher percent population employed in the utilities industry (RRR = 1.038; $p<0.001$) compared to non-host tracts. Attributes for region reveal that living in the West versus living in the South decreases the likelihood of living in a coal host tract compared to a non-host tract (RRR = 0.399; $p<0.001$). Conversely, the likelihood of being located in a coal host tract compared to a non-host tract increases if living in the Midwest (RRR = 4.453; $p<0.001$) or the Northeast (RRR = 2.394; $p<0.001$), compared to living in the South.

Percent population minority, median household income, percent population children, and population density are not significant determinants of being located in a nuclear versus non-host tract. However, tracts with lower percent population poverty are expected to be nuclear host tracts compared to non-host tracts (RRR = 0.980; $p < 0.001$). Additionally, population density is expected to be lower in nuclear host tracts compared to non-host tracts (RRR = 0.878; $p < 0.001$). Further, tracts with greater percent population employed in the utilities sector are more likely to be nuclear host tracts compared to non-host tracts (RRR = 1.092; $p < 0.001$). Dummy variables for region indicate that tracts are less likely to be nuclear host than non-host if located in the West (RRR = 0.159; $p < 0.001$) or Midwest (RRR = 0.702; $p < 0.01$) compared to the South. Conversely, tracts are more likely to be nuclear host than non-host if they are located in the Northeast compared to the South (RRR = 2.361; $p < 0.001$).

Model 1 reveals that both host tracts are less likely than non-host tracts to contain higher percent population minority (RRR = 0.991; $p < 0.05$) and percent families living below the poverty line (RRR = 0.956; $p < 0.001$). Conversely, both host tracts are more likely than non-host tracts to contain higher percent population children (RRR = 1.080; $p < 0.001$) and percent population employed in the utilities sector (RRR = 1.119; $p < 0.001$). Tracts with greater population density are more likely to be non-host tracts than both host tracts (RRR = 0.912; $p < 0.001$). Region variables show that tracts located in the Midwest (RRR = 1.483; $p < 0.05$) and Northeast (RRR = 3.795 L $p < 0.001$) compared to the South, are more likely to be both host tracts than non-host tracts. Median household income, percent population with a bachelor's degree or above and living in the West compared to the South are not significant determinants of being a both host tract versus a non-host tract.

When compared to coal host tracts, nuclear host tracts (see Table 6.3) are more likely to contain populations with lesser percent minorities (RRR = 0.991; $p < 0.001$) and families living below the poverty line (RRR = 0.969; $p < 0.001$). Similarly, tracts with higher median incomes are expected to be located in nuclear host tracts compared to coal host tracts (RRR = 1.009; $p < 0.01$). Interestingly, tracts with higher percent population holding a bachelor's degree are expected to be coal host tracts compared to nuclear host tracts. Compared to coal host tracts, nuclear tracts are expected to be tracts with higher percent population children (RRR = 1.025; $p < 0.05$) and percent population employed in the utilities sector (RRR = 1.051; $p < 0.001$). Additionally, tracts with higher population densities are expected to be coal host tracts versus nuclear host tracts (RRR = 0.865; $p < 0.001$). Variables for region reveal that tracts are more likely to be coal host tracts than nuclear host tracts if located in the West (RRR = 0.399; $p < 0.001$) and Midwest (RRR = 0.158; $p < 0.001$) versus the South. Being located in the Northeast versus the South is not a significant determinant of being a coal host versus nuclear host tract.

Model 2 shows the relationship between coal host tracts and nuclear, and both host tracts. Similarly to nuclear host tracts, both host tracts are less likely than coal host tracts to contain higher percent population minorities (RRR = 0.984; $p < 0.001$), percent families living below the poverty line (RRR = 0.946; $p < 0.001$), and population density (RRR = 0.899; $p < 0.001$). Conversely, both host tracts are more likely than coal host tracts to contain higher percent population children (RRR $= 1.093$; $p < 0.001$) and percent population employed in the utilities sector (RRR $= 1.078$; $p < 0.001$). Regional variables indicate that tracts located in the Midwest compared to the South are more likely to be coal host tracts versus both host tracts (RRR $= 0.333$; $p < 0.001$). In contrast, tracts in the Northeast compared to the South are more likely to be both host tracts than coal host tracts (RRR $= 1.585$; $p < 0.01$). Living in the West versus the South, median household income, and percent population with a bachelor's degree or above are not significant determinants of being a both host tract versus a coal host tract.

Interestingly, when comparing nuclear tracts to both host tracts, percent population minority, percent population with a bachelor's degree, percent population living in poverty, median household income, population density and living in the West compared to the South are not significant determinants of tract type. However, tracts with higher percent population children are expected to be both host tracts compared to nuclear host tracts (RRR $= 1.066$; $p < 0.001$). Additionally, tracts in the Midwest (RRR $= 2.112$; $p < 0.001$) and Northeast (RRR $= 1.607$; $p < 0.01$) compared to the South are also more likely to be both host tracts than nuclear host tracts.

Summary of Multivariate Findings

Coal Host Tracts versus Non-Host Tracts Multinomial logistic regression equations reveal that compared to non-host tracts, coal host tracts contained significantly greater percent of the population that is minority, percent families living in poverty, percent of the population with a bachelor's degree or above, population density, and percent of the population that is employed in the utilities sector. In contrast, coal host tracts had lower median household income and percent of the population that is children compared to non-host tracts. Coal host tracts were more likely than non-host tracts to be located in the Midwest and Northeast versus the South.

Nuclear Host Tracts versus Non-Host Tracts Multivariate analyses reveal that nuclear host tracts possessed lower rates of poverty and population density and higher percent of the population employed in the utilities sector than non-host tracts. Nuclear tracts were less likely than non-host tracts to be located in the Northeast and West. Additionally, nuclear tracts more likely than non-host tracts to be located in the Midwest compared to the South.

Coal Host Tracts versus Nuclear Host Tracts Multinomial logistic regression equations reveal that compared to nuclear host tracts, coal host tracts contained significantly greater percent of the population that is minority, percent families living in poverty, percent of the population with a bachelor's degree or above, and population density, and lower percent of the population that is children. Coal host tracts also had lower median household income and percent of the population that is employed in the utilities sector than nuclear host tracts. Compared to nuclear host tracts, coal host tracts were more likely to be located in the West, Midwest, and Northeast versus the South.

Discussion and Conclusion

This study sought to determine whether race, class, and percent of the population that are children are indicators of power plant injustices. Hypothesis 1 was supported with coal host tracts containing greater percent of the population that is minority than non-host tracts. Further, as predicted in Hypotheses 2 and 3 median incomes were lower and poverty rates were higher in coal host tracts compared to non-host tracts. These findings are congruent with other studies examining inequalities concerning the location of coal-fired power plants in particular (Faber and Krieg 2002; Keating and Davis 2002) and the environmental justice literature more broadly (Bullard 1990; Downey 2005; Mohai and Bryant 1992; Pellow 2000).

Nuclear host tract demographics tell a different story. As predicted in Hypothesis 5, and corresponding to the results of Alldred and Shrader-Frechette (2009), there were no differences in the percent of the population that is minority between nuclear and non-host tracts.

As anticipated, median household incomes were not lower in nuclear host tracts compared to coal host tracts (i.e., Hypothesis 6). Further, as predicted in Hypothesis 7, poverty rates were not greater in nuclear host tracts compared to non-host tracts. These findings were expected as nuclear plants, unlike coal plants, do not appear to have a strong impact on property values in surrounding areas (Farber 1998). Additionally, nuclear host tracts contained significantly lower rates of poverty. Thus, these tracts are not simply unaffected by their host tract status, they appear to be benefitting from decreased rates of poverty. This may suggest that nuclear plants provide economic benefits distinct from other types of hazardous facilities.

Although the aforementioned hypotheses were largely supported by the data, predictions concerning the percent of the population that were children were unexpected. Hypothesis 4 was rejected as there were lower percent of the population that are children in coal host tracts compared to non-host tracts. Hypothesis 8 was rejected with no differences found between percent of the population that are children. These findings are in contrast to previous studies which have indicated children as a group disproportionately exposed to environmental hazards (Grineski et. al. 2010; Perlin et al. 2001).

While it was hypothesized that nuclear host tracts would be more affluent than coal host tracts, given the dangers of radiation exposure associated with living nearby a site, it remains unclear why people with the resources to relocate would select to live near a nuclear powered plant. One possible explanation is that local residents simply do not view nuclear plants as hazardous facilities. The types of additions which result from coal and nuclear plants may help to explain why the former may be viewed as noxious while the latter are not. Emissions from coal-fired power plants can be seen rising from smoke stacks and decreasing visibility, smelt in the air, and felt in the throats and lungs of nearby residents. Community members can make connections in their daily lives between the density of haze in the air and immediate impacts on their local environment and health. For instance, nearby residents may notice increases in coughing, asthma attacks, and exacerbation of respiratory illnesses on days with decreased visibility from coal emissions. Thus, the costs associated with living near a coal plant are undeniable, they can be perceived with human senses and health impacts are both long term and acute.

In contrast, the additions resulting from nuclear powered plants are far more imperceptible and are associated with delayed adverse health effects. The only visible releases from nuclear facilities are clouds of steam. Although escaped radiation has the potential to result in cancer in nearby residents, this exposure cannot be perceived by the affected individual and does not interfere with their daily lives. Additionally, connections between radiation exposure occurring over many years and negative health outcomes such as cancer are not as apparent as suffering from asthma attacks on a day with high levels of smog. Thus, nearby residents of nuclear plants do not experience the same day to day consequences associated with coal plants, although the long-term health impacts may be comparable.

While residents of nuclear host tracts may not be as acutely aware of the emissions resulting from nuclear plants as they might be from coal fired plants, the risks associated with living near a nuclear facility remain well documented. Therefore, it is still not entirely clear why residents with the agency to relocate would select to remain in these hazardous areas. Risk perception theory may offer insight into this perplexity. Risk perception literature informs us that when people receive benefits from a source of risk, they are more likely to selectively overlook information regarding the dangers of that risk (Margolis 1996). If this explains why wealthier communities live in close proximity to nuclear power sites, then we would expect that nuclear plants provide amenities to the community that are not provided by coal-fired plants. One clear advantage of any power facility is the availability of jobs. However, the work environments and number of jobs available to the local community will likely vary by facility type. Jobs provided from construction and operation of coal-fired plants tend to be overestimated, and communities receive little economic benefit from hosting a facility (Ochs Center for Metropolitan Studies 2009). While coal-fired plants provide increasingly limited numbers of jobs to local residents and most employment benefits go outside of the host county, this is not clear in the case of nuclear plant employment. Multivariate analyses from this study support this notion as nuclear and both host tracts possessed greater percent of the population employed in the utilities industry than coal host tracts. While there are other financial advantages of living near a nuclear facility, such as opportunities for trade and improved infrastructure including better road maintenance, the availability of jobs is likely the most influential benefit in terms of risk perception within nuclear plant host communities (Eiser et al. 1995).

It should be noted that this study does not address whether tract demographics were influenced by the presence of power plants, or whether power plants were sited in tracts as a result of their demographic composition. Therefore, this research is unable to answer, for instance, whether certain minorities are disproportionately exposed to coal power plant risks as a result of racist siting decisions or white flight and lower housing costs associated with undesirable facilities. To determine causality, future research should incorporate historic census data and plant construction dates. These methods would offer insight into the types of mechanisms through which power plant inequalities occur. Nevertheless, whether the racism is blatant or institutionalized, the outcome remains the same; minority populations are unequally burdened by the presence of coal power plants in the USA.

This study indicates that power plant risks are not the issue of isolated communities, with more than 20 % of all tracts in the USA falling within 10 miles of a coal-fired or nuclear power plant. While certain populations may be particularly vulnerable at close distances, the hazards from these plants are clearly an issue of national extent. Unfortunately, in our society public health is not prioritized over the objectives of the treadmill of production, which requires a massive supply of economically produced electricity.

The treadmill demands exponentially increasing expansion and requires producers to increase profits in order to survive in the system. Coal and nuclear energy are popular methods of electricity production because they are relatively cheap to produce and thus maximize profits. Coal is particularly appealing in the USA as it is an abundant, and thus inexpensive, domestic resource. Nuclear powered electricity costs even less to produce than electricity generated from burning fossil fuels and is expected to be even cheaper in the future (World Nuclear Association 2012). Producers are further enticed by economic incentives resulting from alliances formed between capital, workers, and the government. Policies advocating energy security and "clean energy" (i.e., energy produced with little to no greenhouse gas emissions) have made nuclear power particularly attractive in the USA. While the negative externalities arising from these forms of electricity production are immense, the treadmill promises to resolve issues of environmental concern including greenhouse emissions, and the safe disposal of nuclear waste through the development of new technologies. Meanwhile, nearby communities exposed to the risks of power plant operations pay a heavy portion of the price of generating low-cost electricity.

Risk society theory informs us that risks from power plants are the expected consequences of our modern lifestyle. However, this does not suggest that these risks are necessary nor that intensive modernization is desirable. Given the large portion of the USA which is potentially impacted by the daily operations of energy production, a reevaluation of the costs and benefits of current energy consumption is needed. Such an analysis must take into consideration the ambiguity in identifying the extent of risks arising from our modern world. It must also consider the players most vulnerable to these risks. While producers may be able to measure the costs of power plant risks in dollar amounts, local communities are vulnerable to the delocalized, non-compensable, and incalculable risks arising from plant operations including long-term contamination of local environments, the loss of life from exposure to emissions, and the unknown health outcomes of a plant accident. Thus, policy makers and government officials responsible for determining acceptable levels of emissions and safety controls should consider not only the cost of the materials necessary for electricity production but also the cost of medical bills and lost work days arising from power plant emissions. If producers, rather than local communities, were responsible for these costs, then perhaps our society would place a higher priority on public health.

References

Alldred, M., & Shrader-Frechette, K. (2009). Environmental injustice in siting nuclear plants. *Environmental Justice, 2*(2), 85–96.

Anderton, D. L., Anderson, A. B., Oakes, M., & Fraser, M. R. (1994). Environmental equity: The demographics of dumping. *Demography, 31*(2), 229–248.

Beck, U. (1992). *Risk society: Towards a new modernity*. Newbury Park: SAGE Publications.

Beck, U. (2006). Living in the world risk society. A hobhouse memorial public lecture given on Wednesday 15 February 2006 at the London School of Economics. *Economy and Society, 35*(3), 329–345.

Brugge, D., & Goble, R. (2002). The history of uranium mining and the Navajo people. *American Journal of Public Health, 92*(9), 1410–1419.

Brulle, R. J., & Pellow, D. N. (2006). Environmental justice: Human health and environmental inequalities. *Annual Review of Public Health, 27*(3), 103–124.

Brunekreef, B., & Holgate, S. (2002). Air pollution and health. *Lancet, 360,* 1233–1242.

Bryant, B., & Mohai, P. (Eds.). (1992). *Race and the Incidence of environmental hazards: A time for discourse*. Boulder: Westview.

Bullard, R. D. (1990). *Dumping in Dixie: Race, class, and environmental quality*. Boulder: Westview Press.

Clapp, R. W., Cobb, S., Chan, C. K., & Walker, B. Jr. (1987). Leukemia near Massachusetts nuclear power plant. *Lancet, 2,* 1324–1325.

Dockery, D. W., Cunningham, J., Damokosh, A. I., Neas, L. M., Spengler, J. D., Koutrakis, P., Ware, J. H., Raizenne, M., & Speizer, F. E. (1996). Health effects of acid aerosols on North American children: Respiratory symptoms. *Environmental Health Perspectives, 104*(5), 500–505.

Downey, L. (2005). The unintended significance of race: Environmental racial inequality in Detroit. *Social Forces, 83*(3), 971–1008.

Eiser, J. R., Van Der Pligt, J., & Spears, R. (1995). *Nuclear neighbourhoods: Community responses to reactor siting*. Exeter: University of Exeter Press.

Energy Action Coalition. (2012). The hub of the youth climate movement. http://www.energyactioncoalition.org/. Accessed 10 Sept 2012.

Energy Justice Network. (2012). Energy justice network. http://www.energyjustice.net/about. Accessed 10 Sept 2012.

Epstein, P. R., & Reinhart, N. (2010). Testimony for the Kentucky General Assembly, House of Representatives Committee on Health and Welfare. London, Ky: Kentuckians for the Commonwealth. http://www.kftc.org/blog/linkeddocuments/documents/Epsteinpercent20Testimony.doc. Accessed 12 Aug 2012.

Evans, S. (2010). Voices from the desecrated places: A Journey to end mountaintop removal mining. *Harvard Environmental Law Review, 34,* 521–576.

Faber, D., & Krieg, E. (2002). Unequal exposure to ecological hazards: Environmental injustices in the Commonwealth of Massachusetts. *Environmental Health Perspectives, 110*(2), 277–288.

Farber, S. (1998). Undesirable facilities and property values: A summary of empirical studies. *Ecological Economics, 24*(1), 1–14.

Galizia, A., & Kinney, P. (1999). Long-term residence in areas of high ozone: Associations with respiratory health in a nationwide sample of nonsmoking young adults. *Environmental Health Perspectives, 107*(8), 675–679.

Gerrard, M. (1994). *Whose backyard, whose risk: Fear and fairness in toxic and nuclear waste siting*. Cambridge: The MIT Press.

Gould, J. M. (1996). *The enemy within: The high cost of living near nuclear reactors: Breast cancer, AIDS, Low birthweights, and other radiation-induced immune deficiency effects*. New York: Four Walls Eight Windows.

Grineski, S. E., Collins, T. W., Aquila, R., de Lourdes, M., & Aldouri, R. (2010). No safe place: Environmental hazards and injustices along Mexico's northern border. *Social Forces, 88*(5), 2241–2266.

Gryparis, A., Forsberg, B., Katsouyanni, K., Analitis, A., Touloumi, G., Schwartz, J., Samoli, E., Medina, S., Anderson, H., Niciu, E., Wichmann, H., Kriz, B., Kosnik, M., Skorkovsky, J., Vonk, J., & Dortbudak, Z. (2004). Acute effects of ozone on mortality from the "Air Pollution and Health: A European Approach" project. *American Journal of Respiratory and Critical Care Medicine, 170,* 1080–1087.

Hendryx, M. (2011). Poverty and mortality disparities in central Appalachia: Mountaintop mining and environmental justice. *Journal of Health Disparities Research and Practice, 4*(3), 50–59.

Hill, B., & Keating, M. (2002). Children at risk: How air pollution from power plants threatens the health of America's children. http://www.catf.us/resources/publications/view/14. Accessed 10 Aug 2012.

Hughes, J. (2006). Uranium mining and milling. *The Ecologist, 36,* 44–47.

Hung, H., & Wang, T. (2011). Determinants and mapping of collective perceptions of technological risk: The case of the second nuclear power plant in Taiwan. *Risk Analysis, 31*(4), 668–683.

Ibald- Mulli, A., Steiber, J., Wichmann, E., Koenig, W., & Peters, A. (2001). Effects of air pollution on blood pressure: A population-based approach. *American Journal of Public Health, 91*(4), 571–577.

Indigenous Environmental Network. (2012). Environmental justice. http://www.ienearth.org/energy.html. Accessed 15 Sept 2012.

Keating, M. (2004). Air of injustice: How air pollution affects the health of hispanics and latinos. http://lulac.org/assets/pdfs/pollutionreport2.pdf. Accessed 15 Aug 2012.

Keating, M., & Davis, F. (2002). Air of injustice: African Americans and power plant pollution. http://www.catf.us/resources/publications/view/20. Accessed 15 Aug 2012.

Long, M., Stretesky, P., Lynch, M., & Fenwick, E. (2012). Crime in the coal industry: Implications for green criminology and treadmill of production. *Organization and Environment, 25*(3), 1–19.

Malin, S. A., & Petrzelka, P. (2010). Left in the dust: Uranium's legacy and victims of mill tailings exposure in Monticello, Utah. *Society & Natural Resources, 23*(12), 1187–1200.

Mangano, J. J. (2002). Infant death and childhood cancer reductions after nuclear plant closings in the United States. *Archives of Environmental Health, 57*(1), 23–31.

Mangano, J. J. (2008). Excess infant mortality after nuclear plant startup in rural Mississippi. *International Journal of Health Services, 38*(2), 277–291.

Margolis, H. (1996). *Dealing with risk: Why the public and the experts disagree on environmental issues*. Chicago: University of Chicago Press.

Markstrom, C. A., & Charley, P. H. (2003). Psychological effects of technological/human caused environmental disasters: Examination of the navajo and uranium. *American Indian & Alaska Native Mental Health Research: The Journal of the National Center, 11*(1), 19–45.

McGinley, P. C. (2004). From pick and shovel to mountaintop removal: Environmental injustice in the Appalachian coalfields. *Environmental Law, 34*(1), 81.

Mohai, P., & Bryant, B. (1992). Race, poverty, and the environment. *EPA Journal, 1,* 6–8.

Mohai, P., & Saha, R. (2006). Reassessing racial and socioeconomic disparities in environmental justice research. *Demography, 43*(2), 383–399.

National Academy of Sciences. (2005). Report in Brief BEIR VII: Health Effects from Exposure to Low Levels of Ionizing Radiation. http://dels-old.nas.edu/dels/rpt_briefs/beir_vii_final.pdf. Accessed 2 April 2012.

Nuclear Energy Agency. (2002). Nuclear energy agency. https://www.oecd-nea.org/. Accessed 15 Nov 2012.

Nuclear Energy Institute. (2012). Nuclear waste disposal. http://www.nei.org/resourcesandstats/documentlibrary/nuclearwastedisposal/factsheet/safelymanagingusednuclearfuel/. Accessed 21 Nov 2012.

Ochs Center for Metropolitan Studies. (2009). An analysis of the economic impacts and financing of the proposed coal-fired power plant in early county: An economic brief by the ochs center for metropolitan studies. http://www.ochscenter.org/documents/EarlyCountyReport3909.pdf. Accessed 14 April 2012.

Pellow, D. N. (2000). Environmental inequality formation: Toward a theory of environmental injustice. *The American Behavioral Scientist, 43*(4), 581–601.

Perlin, S. A., Wong, D., & Sexton, K. (2001). Residential proximity to industrial sources of air pollution: Interrelationships among race, poverty, and age. *Journal of the Air & Waste Management Association, 51*(3), 406–421.

Pilotto, L., Douglas, R., Attewell, R., & Wilson, S. (1997). Respiratory effects associated with indoor nitrogen dioxide exposure in children. *International Journal of Epidemiology, 26*(4), 788–796.

Samoli, E., Aga, E., Touloumi, G., Nisiotis, K., Forsberg, B., Lefranc, A., Pekkanen, J., Wojtyniak, B., Schindler, C., Niciu, E., Brunstein, R., Dodic Fikfak, M., Schwartz, J., & Katsouyanni, K. (2006). Short-term effects of nitrogen dioxide on mortality: An analysis with the APHEA project. *European Respiratory Journal, 27*(6), 1129–1138.

Schnaiberg, A. (1980). *The environment: From surplus to scarcity*. New York: Oxford University Press.

Schnaiberg, A., Pellow, D., & Weinberg, A. (2002). The treadmill of production and the environmental state. In A. P. J. Mol & F. H. Buttel (Eds.), *The environmental state under pressure* (pp. 15–32). London: Elsevier/JAI.

Schneider, C., & Banks, J. (2010). The toll from coal: An updated assessment of death and disease from America's dirtiest energy source. Clean Air Task Force. http://www.catf.us/resources/publications/files/The_Toll_from_Coal.pdf. Accessed 5 Sept 2012.

Sjoberg, L. (2003). Attitudes and risk perceptions of stakeholders in a nuclear waste siting issue. *Risk Analysis, 23*(4), 739–749.

Stone, J. (2001). Risk Perception Mapping and the Fermi II nuclear power plant: Toward an ethnography of social access to public participation in great lakes environmental management. *Environment Science & Policy, 4*(4), 205–217.

Stretesky, P. B., Long, M. A., & Lynch, M. J. (2013). *Treadmill of crime: Political economy and green criminology*. London: Routledge.

Sunyer, J., Ballester, F., Le Terte, A., Atkinson, R., Ayers, J., Forastiere, F., Forsberg, B., Vonk, J., Bisanti, L., Tenias, J., Medina, S., Schwartz, J., & Katsouyanni, K. (2003). The association of daily sulfur dioxide air pollution levels with hospital admissions for cardiovascular diseases in Europe (The Aphea-II study). *European Heart Journal, 24*(8), 752–760.

Swings, P. O., Bowie, C., Philipps, M. J., & Johnson, S. A. N. (1989). Incidence of leukemia in young people in the vicinity of Hinkley Point nuclear power station. *BMJ (Clinical research ed.), 299*, 289–293.

Taliman, V. (1992). Stuck Holding the Nation's Nuclear Waste. *Race, Poverty & Environment Newsletter, 3*(3), 6–9.

Taylor, D. M., & Taylor, S. K. (1997). Environmental uranium and human health. *Review of Environmental Health, 12*(3), 147–157.

United States Census Bureau. (2010). Census tiger/line shapefiles. http://www.census.gov/geo/www/tiger/tgrshp2010/tgrshp2010.html. Accessed 1 April 2012.

United States Energy Information Administration. (2012a). Electricity. http://www.eia.gov/electricity/. Accessed 5 Sept 2012.

United States Energy Information Administration. (2012b). Impacts of coal mining. http://www.eia.gov/energyexplained/index.cfm?page=coalenvironment. Accessed 1 Sept 2012.

United States Energy Information Administration. (2012c). What is the role of coal in the United States? http://www.eia.gov/energy_in_brief/role_coal_us.cfm. Accessed 3 March 2012.

United States Environmental Protection Agency. (2012a). Greenhouse gas emissions. http://www.epa.gov/climatechange/ghgemissions/gases/.html. Accessed 5 Aug 2012.

United States Environmental Protection Agency. (2012b). http://www.epa.gov/. Accessed 8 Aug 2012.

Venables, D., Pidgeon, N., Simmons, P., Henwood, K., & Parkhill, K. (2009). Living with Nuclear Power: A Q-Method Study of Local Community Perceptions. *Risk Analysis, 29*(8), 1089–1104.
World Nuclear Association. (2012). Nuclear power economics and project structuring. http://www.worldnuclear.org/uploadedFiles/org/WNA/Publications/Working_Group_Reports/REPORT_Economics_Report%20(1).pdf. Accessed 7 March 2013.

Ms Sarah Kosmicki graduated with an MSc degree from Oklahoma State University.

Dr. Michael A. Long is a senior lecturer in criminology at Northumbria University in Newcastle, UK.

Chapter 7
Is It a Crime to Produce Ecological Disorganization? Why Green Criminology and Political Economy Matter in the Analysis of Global Ecological Harms

Michael J. Lynch, Michael A. Long, Kimberly L. Barrett and Paul B. Stretesky

Introduction

Historically, orthodox criminology[1] has drawn attention to the concept of the environment in limited ways, preferring to pitch the study of environment as a concept related to control of crimes that occur in public places (e.g. crime prevention through environmental design, or CEPTED; see Jeffery (1977) as an example of this tradition). This more traditional approach to the study of the environment in criminology has also taken the form of hotspot of crime analysis and an emphasis

[1] It is commonplace among radical criminologists to refer to mainstream or traditional forms of criminology—that is, to bureaucratic criminology and its focus on street offenders and their control—as orthodox criminology (see Lynch and Michalowski 2006). This is similar to the distinction made between orthodox and radical economics.

M. A. Long (✉) · P. B. Stretesky
Department of Social Sciences and Languages, Northumbria University, Lipman Building, Newcastle upon Tyne, NE1 8ST, UK
e-mail: michael.long@northumbria.ac.uk

P. B. Stretesky
e-mail: paul.stretesky@northumbria.ac.uk

M. J. Lynch
Department of Criminology, University of South Florida, Tampa, FL, USA
e-mail: mjlynch@usf.edu

K. L. Barrett
Sociology, Anthropology and Criminology Department, Eastern Michigan University, Ypsilanti, MI 48197, USA
e-mail: KLBarrett29@yahoo.com

T. Wyatt (ed.), *Hazardous Waste and Pollution,* DOI 10.1007/978-3-319-18081-6_7

on routine activities theory (Sherman et al. 1989). One recent alternative to that orthodox view of the environment is 'green criminology' that explores green crime and harm from an ecological vantage point, where harms and crimes committed against ecological units and the species living in those units are studied (e.g. Del Olmo 1998; South 1998; Beirne 1999; Stretesky and Lynch 2002, 2011; Walters 2010; White 2011; Long et al. 2012;Westerhuis et al. 2013). Sometimes, these green crimes and harms are examined as direct forms of victimization, such as when an ecological system is damaged by some harmful action (White 2011). In other cases, green criminologists explore indirect forms of victimization, such as species living in ecological areas where a green crime against the ecology is committed (e.g. health harms associated with water pollution). Green criminology also tackles subject matter related to, for example, corporate crimes against the environment that more traditional forms of criminology still tend to overlook (Lynch and Michalowski 2006). Green criminology has made considerable progress in the discipline, but environmental laws and green issues of justice (e.g. environmental justice) are still neglected in criminology. Much of this neglect comes from a lack of material on green criminology as opposed to direct criticism of that material. For example, in their study of 33 years of criminological research published on the topic of environmental justice, Zilney, McGurrin and Zahran (2006, p. 56) discovered that only a few criminologists contribute to the discussion of environmental crime and that 'the number of criminal justice and criminology scholars publishing in the area of environmental justice remain woefully low'. Lynch et al. (2004, p. 293) also discovered that six of the 9410 pages in 16 best-selling criminology textbooks consisted of material that could be classified as dealing with 'toxic waste'-related crimes. Lynch et al. (2004) also reported that, during the mid- to late 1990s, there were no articles published on toxic crime in any of the eight leading criminology journals (Lynch et al. 2004).

When green criminologists engage the criminological literature, they often use a harm perspective to justify why environmental harm, crime, law and justice ought to be examined in the discipline (Del Olmo 1998; White 2011). Criminologists also argue that green harm and crime are more widespread, have more victims and produce more damage than crimes that 'occur on the streets' (Jarrell et al. 2013; Lynch 2013). The major assumption behind the harm approach is that documenting the harm will establish the validity of green criminology by making the empirical case that these harms have substantively significant social and economic consequences, and therefore deserve serious study within criminology (Ellefsen et al. 2012; White 2012; Jarrell et al. 2013; Lynch 2013). Efforts to legitimize green criminology by empirically demonstrating that green crime creates more harm than street crime are noteworthy. Yet, in taking a more practical approach that emphasizes harm, green criminology may diminish the importance of the theoretical rationale for considering green harms[2].

---

[2] We do not diminish the study of street crime. We thank a reviewer for pointing out that similar structural factors are likely related to both street crime and green crime and that both are worthy of study.

In this chapter, we explore the theoretical rationale for a more sustained establishment of green crimes and ecological issues in the discipline of criminology. To do so, we draw on theory and research relevant to ecological disorganization. The ecological disorganization approach draws attention to the ways in which human preferences for organizing economic production consistent with the objectives of capitalism are an inherent contradiction with the health of the ecological system. In that view, capitalism must cause ecological disorganization by consuming and polluting nature. As a result of that inherent contradiction between capitalism and nature, the capitalist system of production must be seen as a crime against nature. The case for that argument is found below. We begin our theoretical analysis by briefly discussing the common approach of identifying the extent of the green crime and harm as it is used to open up space for green criminology in the more traditional literature. Next, we examine the contradictions between capitalism and nature and explore ecological Marxism and treadmill of production (ToP) approaches. We then explore whether ecological disorganization is a crime against nature produced by capitalist systems of production. In the concluding section, we summarize our argument and illustrate how scholars in different fields make similar claims.

The Harm Approach in Green Criminology

Green criminology often draws upon the harm perspective that is broadly conceived by Hillyard and Tombs (2005, p. 7), who suggest that 'there is little doubt that the undue attention given to events which are defined as crimes distracts attention from more serious harm [and] positively excludes them'. As a result, Hillyard and Tombs (2005, 2007) argue that a new social harms approach to criminology is needed if the discipline is to advance scientifically as it lacks ontological reality. In short, the harms approach within criminology would need to be more objective and encompass a variety of physical harms, including 'exposures to various environmental pollutants' (Hillyard and Tombs 2005, p. 10). Hillyard and Tombs point out that identifying harms that people face is clearly difficult, but it is more objective than relying on a body of law that already exists and is influenced by power relations in society. While the researchers see difficulty in identifying the range of harms that exist, they note that this could, in theory, be accomplished within the discipline. Green criminologists have taken those calls by Hillyard and Tombs (2005, 2007) seriously and often rely on a harms perspective in the works they produce (Stretesky et al. 2013b, p. 6). These green criminologists point to the fact that crime statistics often present a distorted view of crime because they fail to include the large volume and scope of crime and harm that ecological disorganization produces. In short, green criminologists often refer this harm in comparison to traditional 'street' crimes that the state records for statistical purposes for crimes, such as murder, rape, robbery, assault, larceny and motor vehicle theft (Jarrell et al. 2013). Lynch (2013) has recently noted that green crimes easily surpass the volume and number of victims reported in crime statistics that are kept by the state. As Hillyard and Tombs (2005, 2007) have argued, there are a wide variety of green crimes, and the victims of green crimes include nontraditional victims whom criminologists do not ordinarily examine (Hall 2013). In addition to human victims, green crimes also have nonhuman victims including animals, plants and ecosystems (Ellefsen et al. 2012; Nurse 2013). Green crimes do not outnumber the crimes reported by the police, such as murder, rape, assault, larceny, burglary and motor vehicle theft simply because there are more categories of victims (Lynch 2013). For instance, green harms often victimize larger numbers of human victims in a single incident compared with typical street crimes (Williams 1996b). A single green crime may produce hundreds, thousands or even millions of human victims (Hillyard and Tombs 2005; White 2012). Some of those victims suffer repeated victimization as green crimes can also unfold over long periods of time and have a duration not typically associated with street crimes (Williams 1996a, b). Each of these factors increases the scope, intensity and numbers of green victimizations, making these forms of victimization quite different from the typical street crime victimization incident (Williams 1996a; Hall 2013; Lynch 2013). Green harm and crime are conceptually as well as theoretically important because they have the ability to cause forms of ecological damage that change the very nature of the world (White 2012). These green harms can also make the world uninhabitable. Abandoned towns and communities exist because of the health hazards posed by toxic pollutants and other related environmental disasters that are counted among green crimes (Lynch 2013). In the USA, for example, these locations include the following: Times Beach, Missouri (due to dioxin pollution); Centralia, Pennsylvania (due to underground mining fires); Love Canal, Niagara Falls, NY (due to widespread disposal of toxic waste); Pitcher, Oklahoma (due to concentrations of lead and zinc pollution); Treece, Kansas (due to lead pollution). These cities, and others around the world, stand as monuments to the tremendous harms that green crimes can produce. In addition to these abandoned cities, there are currently 1163 superfund sites that are portions of cities, towns and communities in the USA listed as containing sufficient levels of ground pollution to require remediation (EPA 2013b).

It should also be noted that the financial costs of exposure to environmental pollutants are substantial (Hillyard and Tombs 2005). For children, Landrigan et al. (2002) estimate the cost of lead poisoning in the USA at US $ 43.3 billion. For three other environmental pollutant outcomes (asthma, cancer and developmental disabilities), the cost for children's health care was estimated to amount to as much as an additional US $ 21.4 billion (Landrigan et al. 2002). In this example, we can see that the one-time annual costs for US children, who comprise about 24 % of the population, for only four environmental health outcomes are considerable (McCollister et al. 2010). Thus, if we consider that there are other disease outcomes associated with exposure to environmental toxins and pollutants, and that there are populations excluded from the above estimates (i.e. adults), the costs of green crimes on human health is likely to be excessive.

It is instructive to note that several important trends related to green crimes affect ecological systems locally and globally. In the USA, for example, forest land has become more fragmented, resulting in an overall loss in forest cover (Griffith et al. 2003). This negatively impacts species maintenance and the health of the ecosystems, particularly in relation to climate and temperature change (Fall et al. 2010). Similar results are found for the global ecosystem and its stability (on biodiversity, see Butchart et al. (2004); on carbon and deforestation, see Cramer et al. (2004); on wetlands, see Zedler and Kercher (2005); on wetland loss due to rising sea level, see Nicholls et al. (1999)).

Theoretically, there are important differences between green criminology and traditional forms of criminology. The traditional criminologists draw attention to the individual and their character and relationships as the salient features of crime (Hillyard and Tombs 2005). In the green literature, however, the focus is drawn to how structural forces such as capitalism and mechanisms within capitalism, such as the ToP generate ecological crimes by the very nature of the structural composition of these forces of production (see Long et al. 2012; Stretesky et al. 2013a, p. 6). In the terms of ecological Marxism, this means that the structures of capitalism and nature are in inherent contradiction and conflict, with each expansion of capitalism promoting ecological destruction and the disorganization of nature (see below and Foster 2000; Burkett 2006).

What the above indicators describe is a dramatic change in both local and global ecosystems produced by green harm. Theoretically, the question is, what ties these ecological changes and harms together in a coherent and useful way? In general terms, these problems relate to ideas of change and transition found within the major macro-level school of criminology: social disorganization theory. Social disorganization theory suggests that major changes in urban areas are the result of large-scale social transitions, such as industrialization and urbanization. Geographers have long traced the trajectories of industrialization and urbanization to the growth of capitalism (Scott 1986; Bensel 2000). They have treated industrialization and urbanization in a political economic context—an approach that has largely been ignored within the criminological literature. In cross-cultural criminological research, these effects (industrialization/ urbanization) have been depicted as measures of modernization (e.g. Bennett 1991), and much cross-cultural research addresses the effects of modernization on crime (e.g. Austin and Kim 1999). Similar research can also be found with respect to the effects of industrialization, urbanization and modernization on ecological issues (e.g. Ehrhardt-Martinez et al. 2002).

Ecological researchers have also provided a broader explanation that ties industrialization, urbanization and modernization together to understand the emergence of ecological problems. Based on political economy, this explanation posits that capitalism precedes industrialization, urbanization and modernization in contemporary contexts. That approach for explaining ecological disorganization has been employed widely in the ecological and economics literatures (Humphreys 2003; Clark et al. 2010; Jorgenson 2010; Jorgenson et al. 2010; Mbatu 2010; Jorgenson and Clark 2011; York et al. 2011; Burgess et al. 2012; Clark and Jorgenson 2012; Clark et al. 2012; Jorgenson et al. 2012; Long et al. 2012).

Within that literature, both ecological Marxism and the ToP theory use political economic theory to demonstrate that there is an inherent conflict or contradiction between capitalism and nature. That inherent conflict means that capitalism must destroy nature to advance (Foster 2000; Burkett 2006). It is to that view that we turn our attention.

The Contradiction Between Capitalism and Nature

Two environmental theories that are sometimes joined together—ToP theory and ecological Marxism—argue that capitalism and nature are in contradiction with one another (Schnaiberg 1980; Foster 2000, 2012; Burkett 2006; Foster and Burkett 2008). To understand the logic of this argument, it is necessary to outline the basic operational principles of capitalism, and then to extend those principles to a discussion of ecological disorganization.

As a system of production, capitalism is based on unequal ownership of the means of production (Marx 1974). The owners of the means of production must employ labour to operate the means of production. That labour is provided by the working class who, under capitalism, has no alternative but to offer its labour for sale to the owning class. The owners of the means of production organize production in such a way that labour is exploited to generate surplus value. Surplus value is the value above production costs that is retained by the capitalist class. In essence, the volume of surplus value retained by the capitalist class is a function of the exploitation of labour in the production process.

Capitalism has one widely agreed-upon goal—to increase the volume of the profit produced. In Marxian analysis, the amount of profit is partially determined by the rate of surplus value and expanding production—that is, by increasing the volume of goods produced and sold (Marx 1974). Thus, in order for profit to expand, both the production process and the exploitation of labour must constantly increase.

Marx analysed this process in the production phase and assumed that the input for production—raw materials and capital—were prerequisites for the initiation of production. However, it was not Marx himself who exposed the inherent contradiction between capitalism and nature, but rather ecological Marxists who, drawing on Marx's earlier writings, initiated the discussion of this contradiction (Foster 2000; Burkett 2006). In doing so, ecological Marxists illustrated that, in order for production to begin and expand, inputs into the system must be available and also must expand. In other words, for capitalism to expand, it must increasingly consume larger quantities of natural resources. The expansion of capitalism, therefore, leads to the escalation of ecological destruction through consumption of nature's raw materials in the production process.

In contrast to the capitalist system of expanding resource consumption, nature's productive system is based on growth through conservation (Foster 2000; Burkett 2006). Nature has its own system of production (Foster 2000; Burkett 2006) where it produces goods sustainably and promotes the development of the forms of life existing on the Earth (Kovel 2007). The disruption of that system causes various forms of environmental instability. One main example in the modern era is climate change (Lovelock 2000, 2007). Other forms of environmental instability include the following: the increase in pollution; pollution's changing forms, locations and scope; the destruction of forests and wetlands; the loss of plant and animal diversity; and so forth.

The Treadmill of Production

Schnaiberg (1980) introduced his theory on the ToP that explored the capitalism–nature contradiction from the perspective of the consequences of production. Schnaiberg's ToP perspective introduced the concept of ecological disorganization, and illustrated the ways in which the capitalist ToP produces ecological disorganization.

Schnaiberg chose the phrase 'treadmill of production' as a descriptor of capitalism that was consistent with the argument that capitalism must constantly expand and is constantly in motion. The capitalist system is constantly moving and producing to facilitate its inherent growth imperatives. Schnaiberg's argument also suggested that the capitalist ToP changed significantly following the Second World War. After an initial period of expansion following the Second World War, human labour inputs into the system of production declined, especially relative to the forms of nonhuman labour that the system was increasingly incorporating. In Marxian economic terms, the organic composition of capital was radically transformed following the Second World War, as capital costs were shifted from labour to machinery and, more importantly, towards chemically and energy-intensive forms of nonhuman labour. Several consequences followed these transitions.

Ecological Additions

ToP approaches address ecological disorganization associated with ecological additions. Ecological additions consist of pollutants that the ToP adds to the environment. Following the Second World War, the nature of those pollutants began to change, whereas the number of ecological additions relative to production rose as treadmill practices turned towards the use of chemical and energy-intensive production practices. As the ToP increased the use of chemicals in production to intensify and replace human labour, it began to produce large quantities of chemical pollution. The release of energy-related pollution followed suit as the treadmill consumed larger and larger quantities of fossil fuels. The extent of these processes over time cannot be measured completely, since data on the volume of pollutants produced by the ToP were not kept, at least in the USA, until 1988, and data reliability issues for the first few years of the data collection process resulted in underreporting. In 1991, however, under the Toxic Release Inventory (TRI), the US Environmental Protection Agency (EPA) reported that the 25,081 reporting facilities generated more than 29 billion pounds of waste, or about 1.16 million pounds of waste per facility. The quantity of TRI waste produced rose to nearly 34 billion pounds annually by 2000 (1.4 million pounds per facility), then it declined through 2010 during the long economic recession in the USA to about 20.4 billion pounds (950,000 pounds per facility), and finally increased to 22.8 billion pounds of waste (1.09 million pounds per facility) by 2011. In the first decade of the twenty-first century, TRI reporting facilities, which are only a small portion of the facilities the EPA regulates, reported producing more than 239 billion pounds of toxic waste. Research suggests that these TRI figures underestimate the total waste produced because of underreporting (de Marchi and Hamilton 2006) and changes in TRI reporting requirements (Bennear 2008). The total waste produced may be up to 40 % higher than reported.

Ecological Withdrawals

Along with ecological additions, a second major issue that ToP theory addresses is the effect of the ToP on ecological disorganization through ecological withdrawals. Ecological withdrawals include the collection of raw material input into the treadmill and the forms of ecological damage created in accessing those materials. These withdrawal effects include operations such as the following: the mining of minerals, coals, ferrous and nonferrous metals, chemicals and precious stones; the drilling for oil, natural gas and other fossil fuel sources; and the harvesting of wood. These effects can also include the consequences of withdrawal, such as the impact of changing the landscape through mining on ecosystems, or the effects caused by withdrawing a significant volume of water to facilitate withdrawal processes, or the pollution of land, air and water associated with withdrawal processes.

Ecological withdrawals have important impacts on ecological disorganization. As an example of the widespread impacts that can be caused by one form of ecological withdrawal, we provide some detail concerning the impacts of timber harvesting and deforestation on local and global ecosystems. These include, namely, local climate change (Nobre et al. 1991; Zhang et al. 1996); changes in local precipitation (Kanae et al. 2001; Negri et al. 2004); changes to the global carbon cycle (Cramer et al. 2004); impacts on other nearby types of forests lands (e.g. the effect of tropical lowland deforestation on montane cloud forests, Lawton et al. 2001; Ray et al. 2006); impacts on tree species diversity (Hubbell et al. 2008) and genetic diversity of tree stock (Lowe et al. 2005; Sebbenn et al. 2008); forest re-growth cycles (Zou et al. 1995); changes in microbial soil diversity (Borneman and Triplett 1997); effects on soil erosion (Van Rompaey et al. 2002) and soil nutrient run-off (Zheng et al. 2005); impacts on animal species diversity (Harvey et al. 2006), fish species diversity (Lorion and Kennedy 2009) and invertebrate diversity (Harding 2003); and numerous impacts on various natural water systems (Sweeney et al. 2004; Coe et al. 2009). There are, of course, damaging impacts caused by other ecological withdrawal processes such as, mountaintop removal mining; mining of heavy metals; hydrofracking; mineral mining; oil drilling; coal tar, coal shale and coal sand extraction; mining of nuclear materials; etc. Clearly, withdrawals that result from the ToP cause extensive harm.

The process of ecological withdrawal not only harms the environment directly through extraction methods but also contributes to the production of ecological additions when certain extraction methods are employed. For example, coal processing generates significant quantities of coal sludge during the cleaning process and the preparation of extracted coal for the market. This process uses large quantities of water that can draw down local water tables. The coal slurry that results from preparing coal for market contains a number of hazardous chemicals (more than 60 are used). Coal sludge can also contain heavy metals that coexist in coal formations, including mercury, arsenic, beryllium, cadmium, nickel and selenium. These wastes are often kept in coal impoundments, which themselves are often built from the coal waste (for additional information, see the Coal Impoundment Location & Information System (CIL&IS), www.coalimpoundment.org). The CIL&IS system lists 307 coal impoundments in the main coal-producing states of West Virginia, Kentucky, Ohio, Pennsylvania and Tennessee. Human health often suffers from the impact of ecological withdrawals associated with coal mining. Health-related quality-of-life indicators are significantly worse for the US counties that contain coal-mining operations compared to those that do not contain coal-mining operations (Zullig and Hendryx 2010).

Synthesis

In the preceding sections, we addressed the contradictions between capitalism and nature, pointing out that ecological Marxism and ToP approaches contain theoretical tools for laying out the character of that contradiction. To summarize that view, the inherent expansionist logic of capitalism, which requires constantly expanding consumption of raw materials found in nature, undermines nature's ability to reproduce conditions for life. In addition, with the help of human labour, capital transforms nature's labour into objects that contain value in the human economy. Many of those objects, and the capital exchanged for those objects, are luxury items—items that are not necessary for human existence, but rather are items of pleasure. Those luxuries come with a significant cost—ecological disorganization. Therefore, those individuals or corporations that amass the greatest concentrations of wealth are also those likely to have the greatest impacts on the disorganization of nature. In other words, those with the most accumulated wealth have the ability to use their wealth to consume considerably more of nature's wealth than others.

These observations raise two questions, one of which we shall not attempt to address here. The unaddressed question has to do with how much can be consumed, and a related question of what is fair to consume. Those questions are beyond the scope of the present argument, but are valid questions for green criminologists to address and have important implications for criminological issues of justice. The question we will address in the section that follows is whether the ecological destructive consequences of capitalism are sufficient to say that capitalism is an organized form of economic activity that produces crimes of ecological disorganization[3].

Is Ecological Disorganization a Crime of Capitalism?

In the orthodox tradition, a crime is a violation of the law, a behaviour reprehensible enough to call forward the condemnation of society in its written rules. The assumption behind that approach to defining crime is that the law is a useful means for discerning the difference between behaviours that are and are not crime. It is our position, however, that the law is not a useful mechanism for this purpose; the legal definition of crime is not an objective, scientifically derived definition. Legally, crime is simply a behaviour that the law labels as such. There is no standard in the law that describes the behaviours that ought to be crimes—no objective, independent definition of crime outside the labels the law creates and applies. As a social construction (Quinney 1970), crime is flexible. This is particularly true where environmental matters are concerned, as those behaviours that should be considered crimes against the environment have been hotly debated (see, e.g. Beirne 1999, 2009).

In the case of behaviours that ought to be labelled green/environmental crime, we favour scientific definitions of harm over the socially constructed definitions criminologists use when they rely on legal definitions of crime. Of particular concern is that socially constructed legal definitions of environmental/green crimes vary widely and appear in different forms of law and even international treaties (White 2011). This makes green crime very difficult to define. Environmental law allows many harmful behaviours to escape regulation because political and social interest groups strongly influence environmental law decision-making practices.

We posit that a green crime is a behaviour that produces unnecessary ecological harm—harms that can be avoided by organizing production in different ways than are currently practised. Not only are these green crimes unnecessary, they

[3] A reviewer raised a concern about our use of the word 'crime' in the context of ecological disorganization. We use both 'crime' and 'harms' to refer to the effects of ecological disorganization. Hillyard and Tombs (2005, 2007) make the case that researchers should keep the concepts of crime and harm separate. We include both the terms together to indicate that ecological disorganization creates harm and is sometimes defined as criminal. We also believe that referring to ecological and environmental 'crime' is important because it helps make the case that studies of this type should be within the purview of the discipline of criminology. We expand on the discussion of ecological harm and crime in the discussion section.

promote the disorganization of nature. Furthermore, the harm that is imposed on the environment can be concretely defined by scientific standards. To illustrate the points we have made about the definition of green crimes, we provide an example of the potential problem of not using scientific standards of harm to define green crime.

Our example involves differential definitions of environmental harms related to lead exposure/pollution. While there is extensive agreement in the scientific literature that there is no safe level for exposure to lead (Pb), the law permits Pb to be emitted into the environment and in the workplace at certain levels (California Department of Education 2013; EPA 2013a; Protection of Environment 40 C.F.R. § 80.2). The exposure levels the law contains are not scientifically agreed-upon standards, since scientists agree there is no safe level of exposure to Pb (Gilbert and Weiss 2006). However, the environmental rules regarding Pb exposure that have been constructed reflect economic interests, privileging economic actors over public health and scientific standards for exposure (Rosner and Markowitz 1985; Needleman 1997; Markowitz and Rosner 2003; Wedeen 2009).

Employing our definition of green crime detailed above, ecological disorganization is a crime when the behaviour in question produces forms of ecological disorganization that nature cannot accommodate and science can identify as a harm. In the modern capitalist world economy, it may be economically beneficial to extract resources from countries with low wage rates, ship them to other low-wage-rate nations to transform them into commodities and then ship them to other nations for consumption (Gereffi 1999). While this behaviour is beneficial for capitalism in terms of expansion and profit, it is not beneficial for the environment. The environmental impacts of the international capitalist economy include escalated carbon dioxide pollution (Li and Hewitt 2008; Stretesky and Lynch 2009) and other hazardous pollutants (Matthews et al. 2001). As some researchers note (Matthews et al. 2001), corporations fail to consider how their behaviours impact ecological disorganization because they externalize the costs of ecological harms—ecological problems that are produced by corporations become social problems which the government, rather than the private firms that create those problems, must address. This shifts the expense of anti-environmental practices to the state, which must tax citizens to generate the funds for remedies. Corporations benefit from the combination of weak environmental regulations and a pseudo-free market that enables the corporations to externalize costs to the state. This in turn facilitates private-sector accumulation by transferring corporate costs to individual tax payers (see, e.g. O'Connor 1973).

In short, pollution is a cost of production in the contemporary system of capitalism. The forms of pollution that currently exist have been shaped by capitalism and the ToP. The capitalist ToP maintains low commodity prices by taking advantage of differentials in the costs of labour in the international labour market and by using chemically and energy-intensive production practices that generate significant ecological disorganization through ecological withdrawals and additions (Schnaiberg 1980). Together, these treadmill practices help push the expansion of production and consumption forward, continually escalating global ecological disorganization. In contemporary capitalism, corporations are able to externalize productive costs associated with ecological disorganization because they are not required to address those costs in any substantial way during the production and distribution of commodities. In the modern world, the cost of responding to ecological disorganization is borne by governments. Governments have a series of decisions to make about how to best protect capitalism, nature and the citizens they represent. Should a government charge the polluters fees for polluting? What are the economic consequences of doing so when other governments do not make that same choice? If decisions about the control of ecological disorganization are not reached by governments in a unified way through, for example, treaties, then global corporations can simply avoid social control by moving their productive forces to a new location.

The recognition that global capitalism cannot be effectively controlled by state-specific regulations makes it clear that it is difficult to deter corporations from producing ecological disorganization through traditional market mechanisms (Stretesky et al. 2013a). Small firms often lack sufficient capital to escape regulation through transferring production, giving large corporations an advantage that over time leads to the concentration of capital in large global corporations. As a result, the large corporations acquire an increased capacity to avoid regulations in the fragmented regulatory environment of the contemporary world where ecological regulations vary from nation to nation. For state social control of ecological disorganization produced by global corporations to be effective, it is not so much the size of fines that matter as much as their uniformity across nations[4]. Until national governments recognize that they must become global in the same way as capitalism, they will be ineffective when they attempt to control the modern, global capitalist ToP and the resulting ecological disorganization (Gould et al. 2008). Global corporations will select production policies that provide the greatest return to shareholders without regard for the environmental consequences[5] or the impacts its behaviour has on public health, unless external influences require the global corporation to do so.

Consider, for example, maritime shipping. Maritime shipping offers the ability to move large quantities of product over long distances where less expensive modes of transport (e.g. rail) are unavailable. While cost conscious for international trade, maritime shipping has significant ecological consequences. It is estimated that maritime shipping emits more than 1 billion tons of atmospheric carbon dioxide annually (Linstad et al. 2011). Studies indicate that carbon dioxide reductions of between 19 and 28 % can be achieved by regulating the maximum speed at which shipping vessels are allowed to travel (Linstad et al. 2011). This form of social control has no costs to corporations other than extending the amount of time a shipment takes, while providing significant positive ecological benefits. To create such regulations effectively, all nations would need to impose maximum speed regulations in waterways they control—an outcome that would be possible

[4] The same could be said about political divisions within nations.

[5] Exceptions to this rule exist.

through international maritime regulation. Will corporations see the wisdom in these efforts to protect the environment? Or will they lobby governments to prevent this kind of regulation? In general, corporations are not concerned with the external, ecological costs and benefits of their actions or the long-term consequences of their behaviour—only with how their present activities impact profit in the short run. With respect to merchant shipping, for example, studies indicate that climate change is likely to lower water levels in the Great Lakes, meaning that vessels that operate in the Great Lakes will be required to reduce their loads, resulting in the need for more shipments (Millerd 2011) at increased environmental costs, such as expansion of greenhouse gas emissions. In this case, climate change is expected to affect future profits, leading to a reduction of 5–22 % in profits, depending on the extent of climate change (Millerd 2011). Those future costs, however, are not a concern to corporations that base their economic analysis on short-term data and returns and, in doing so, are not considering the external costs associated with climate change. Capitalism is not concerned with what the world will look like in 25, 100 or 1000 years as it continues to churn the ToP, consuming nature and reconfiguring it into commodities for the sole purpose of capital accumulation.

Discussion and Conclusion

Capitalism constantly destroys and disorganizes nature as it seeks to expand at any cost. As Ernest Mandel (1990, p. 23) wrote, 'The capitalist's compulsion is to accumulate'. As for accumulated capital, it is employed continuously in search of more profit, to expand itself indefinitely, which is impossible according to the rules of physics because you cannot create something (e.g. accumulated capital) out of nothing. Accumulated capital is the end result of capitalism's transformation of nature's wealth.

We can, of course, turn to Marx for a more in-depth analysis. As Marx wrote in Capital:

> Accumulate! Accumulate! That is the Moses and the prophets! … Accumulate for accumulation's sake, production for production's sake: by this formula classical economy expressed the historical mission of the bourgeoisie, and did not for a single moment deceive itself over the birth-throes of wealth. (Marx 1974, p. 595)

This is the not-so-subtle secret of capitalism—to grow continuously, to accumulate and to ignore anything outside of accumulation and expansion.

Others have written eloquently about the problem of ecological destruction more generally, though not necessarily from economic or Marxian perspectives. In her seminal work, Silent Spring, noted biologist, Rachel Carson (1962) wrote:

> Only within the moment of time represented by the present century has one species—man—acquired significant power to alter the nature of the world. During the past quarter century this power has not only increased to one of disturbing magnitude but it has changed in character. The most alarming of all man's assaults upon the environment is the contami-

> nation of air, earth's rivers and seas with dangerous and even lethal materials. This pollution is for the most part irrecoverable; the chain of evil it initiates not only in the world that must support life but in living tissues is for the most part irreversible. In this now universal contamination of the environment, chemicals are the sinister and little-recognized partners of radiation in changing the very nature of the world…. It took hundreds of millions of years to produce the life that now inhabits the earth—eons of time…. Given time— time not in years but millennia—life adjusts … but in the modern world there is no time. (Carson 1962, pp. 5–6)

Silent Spring should have marked a historical transformation in societies around the world—especially industrial societies, since these nations bear the largest responsibility for industrial crimes against nature. Yet, that moment in time passed without much change (Kovel 2007). For, if it had, the second coming of Silent Spring, which we argue was ushered in by Colborn et al. (1996) book, Our Stolen Future, and its message about the effects of hormone-disrupting chemicals on the environment and the species of the world would not have been necessary.

In another groundbreaking work, biologist Sandra Steingraber (1997) examined the relationship between the industrial chemical pollution and cancer. While she laid out the scientific evidence for the connection, she also pointed out that governments contributed to the current chemical blizzard by establishing ineffective regulations. For example, despite the fact that numerous studies indicate the 'exquisite vulnerability of the fetus to cancer-causing chemicals', regulatory standards for exposure in the USA and Britain are based on adults, ignoring the foetus or the health of pregnant women (Steingraber 1997, p. 278). Steingraber argued that, at this point in history, it was time to take a human rights approach to cancer. While an important observation, in our view is that this is not just a human rights concern, but a concern for all species, and a concern best framed in relation to the rights of nature as a living being. It is not just chemical pollution from which we must protect the Earth and its species, but from capitalism and the ways in which it employs chemicals in the ToP to intensify the destruction of nature through ecological withdrawals and additions.

Capitalism's unending desire to accumulate and its ecologically destructive forces are serious crimes. Capitalism is based on the exploitation of nature and human labour, and the unequal distribution of the results of exploitation. It does not seek justice; rather it is simply in search of more. Many books have been written on the destructive consequences of capitalism (e.g. Foster et al. 2010; Magdoff and Foster 2011). Some of these stories are haunting, such as Erik Reece's (2006) ethnography of Lost Mountain and its surrounding communities and ecosystems. This is the story of a mountain being destroyed by mountaintop removal mining. It is also the story of how capitalism, in its search for efficiency, uses the ToP to create crimes of ecological disorganization.

Those who have chronicled the destruction of the Earth—including eminent scientists who study extraordinary problems like climate change—tend not to place blame on capitalism. However, we argue that the ecological problems of the modern era cannot be understood without the backdrop of capitalism, the world's driving economic form, which has transformed itself since its emergence in the short world time of five centuries. Primates first appeared on the Earth about 80 million years ago. The 'great apes', including humans, made their first appearance

15 million years ago. It is only in the last few 100 years, a period that exists wholly within the domination of the capitalist world system, that ecological destruction has become so intensified that it threatens the existence of the world. The history of the crime of ecological disorganization cannot be disentangled from the history of capitalism.

This claim is evident, as we argued above, in the work of ecological Marxists and ToP theorists and researchers. Criminologists, and in particular green criminologists, seem to shy away from connecting capitalism to ecological crime. If green criminology is to mature, to move beyond the case study analysis and undercover the causes of ecological destruction, it must do so with the aid of political economic analysis.

Green criminology is revolutionary to the extent that it has opened up space for the discussion of green harms and crimes—a space that, before the twenty-first century, was small within the criminological literature. The initial era of that revolution in criminological thinking has occurred. In the past decade, green criminology has successfully made some inroads into the criminological discipline, and one can find some articles scattered in journals on green crime and a book series on green criminology now exists. Moreover, the green criminologists of the world have organized the International Green Criminology Working Group to help solidify their place in the discipline. These are important steps forward. This does not mean, however, that green criminology is widely recognized in the discipline. One does not find, for example, articles on green criminology published in the high-ranking journals in the discipline, government agencies rarely provide funds for the study of green crimes, and green crime is not regularly addressed in criminology textbooks.

Despite significant news about the expanded state of environmental harm in the world and the expanding scientific literature on that point, criminologists have been slow to take up these kinds of concerns (Agnew 2012). For example, among traditional criminologists, almost no attention has been directed to issues, such as climate change (Agnew 2012), whereas green criminologists have made significant reference to this issue (Stretesky and Lynch 2009; Lynch et al. 2010; Lynch and Stretesky 2010; and chapters in White 2012).

We would also like to clarify our position on the definition of crime, science, ecological crime and discussions concerning the forms of ecological harm that ought to be considered crimes. First, with respect to the definition of crime, criminology has been marked by two primary opposing positions. The first defines crime as a violation of law, and treats the violation of law as if it were an objective standard for defining crime. The second view is that crime is a social construction and hence that it is a highly subjective matter. None of the views, in our opinion, is totally correct. To be sure, when these views are combined, they tell us something about how crime is created and, in particular, the relationship between crime and law—that crime is subjectively constructed in law through bureaucratic regimes that make the construction of law appear objective. There is, however, nothing objective about creating crime in this way, and to view crime as wholly subjective makes little sense either if there is ever a hope that law escapes its tendency to reinforce capitalist relations of production.

In the ecological disorganization view we take, neither the criminal law nor the constructionist view of law is adequate. That is because, when the ecological system is disorganized by human activity, the harm that emerges is measureable. For example, when humans pollute the environment, the level of pollution can be measured and its harm can be detailed in studies showing the impact of exposure to those toxins by various species and the ecosystem itself. Scientists can measure these outcomes, and do, using comparisons of pollution to background chemicals in the environment, and other scientific measures such as the anthropogenic enrichment factor (AEF). Scientists can measure changes in disease rates as pollution is added to the environment; they have the ability to measure changes in air quality in different ways and for different purposes, can measure climate change, and many other conditions harmful to the ecosystem and the species that inhabit ecosystems. In this sense, the crime of ecological disorganization is objective, much more so than criminal law or social constructionist perspectives.

Second, for the criminologist, this raises a question posed perhaps more from resistance to efforts to change the discipline of criminology than anything else. That question states: is it possible to measure all forms of ecological harms objectively? At the present juncture in history, criminologists do not know the answer to that question, and have not tried to address this issue. Ecological scientists, however, have brought science to bear on many of these concerns. To be sure, they are unlikely to have addressed all the ways in which ecological harms can be measured objectively, but that does not mean that they have not built up an impressive catalogue of evidence as to how this can be done.

Third, in reaction to this suggestion, the criminologist may reply: is it really an ecological harm/crime to kill a something as small and seemingly insignificant as a mosquito? The trouble with this question is that it frames the problem at the wrong level of analysis and hence misses the point of examining how humans produce ecological disorganization. Humans do not kill one mosquito at a time—they kill them by the hundreds, thousands and millions by liberally applying pesticides to the environment. That liberal application of pesticides kills other species and changes how nature works, often leaving long-lasting toxins in the environment. In the contemporary period, structural ecological harms are continually aggregated, which causes severe ecological disorganization and has a great impact on the organization and health of nature. These harms should no more be posed as micro-level questions (killing a mosquito) than should the social problem of street crime. Both, in our view, have to do with the nature of the modern system of world capitalism.

In closing, as Joel Lovelock (2007, p. 3) notes, in the era of environmental awareness that began in 1970, the ecological state of the world has gotten worse. As he states, the era of environmental awareness has 'been the era of greatest environmental breakdown'. Kovel argues that we face a choice: the end of capitalism, or the end of the world. According to Kovel, we are now 'capitalism's puppet', acting according to its laws, letting it rule us instead of things being the other way around. He concludes that capitalism violates 'the nature of nature' and human nature, and needs to be replaced by a way of life that promotes the health of the planet, humans and other species.

Until green criminology can become radical enough to take such a stance, it has done little more than open an academic space where a small portion of criminologists can call attention to the ecological ills and harms that routinely occur around us. The solution to these problems is not to merely say 'Ah, here is something criminologists ought to study', but to identify the forces that drive that process so they can be remediated. The problem is capitalism and crimes of ecological disorganization. The sooner green criminologists recognize this, the better equipped they will be to understand the real problem and what must be done.

References

Agnew, R. (2012). Dire forecast: A theoretical model of the impact of climate change on crime. *Theoretical Criminology, 16,* 21–42.

Austin, R., & Kim, Y. (1999). Educational development and homicide in Sub-Saharan Africa. *Homicide Studies, 3,* 78–98.

Beirne, P. (1999). For a nonspeciesist criminology: Animal abuse as an object of study. *Criminology, 37,* 117–148.

Beirne, P. (2009). *Confronting animal abuse: Law, criminology, and human-animal relationships.* Lanham: Rowman & Littlefield Publishers.

Bennear, L. (2008). What do we really know: The effect of reporting thresholds on inference using environmental right-to-know data. *Regulation and Governance, 2,* 293–315.

Bennett, R. (1991). Development and crime: A cross-national time series analysis of competing models. *Sociological Quarterly, 32,* 343–363.

Bensel, R. (2000). *The political economy of American industrialization, 1877–1900.* Cambridge: Cambridge University Press.

Borneman, J., & Triplett, E. (1997). Molecular microbial diversity in soils from eastern Amazonia: Evidence for unusual microorganisms and microbial population shifts associated with deforestation. *Applied and Environmental Microbiology, 63,* 2647–2653.

Burgess, R., Hansen, M., Olken, B., Potapov, P., & Sieber, S. (2012). The political economy of deforestation in the tropics. *The Quarterly Journal of Economics, 127,* 1707–1754.

Burkett, P. (2006). *Marxism and ecological economics: Toward a red and green political economy.* Chicago: Haymarket Books.

Butchart, S., Stattersfield, A., Bennun, L., Shutes, S., Akçakaya, H., Baillie, J., Stuart, S., Hilton-Taylor, C., & Mace, G. (2004). Measuring global trends in the status of biodiversity: Red list indices for birds. *PLoS Biology, 2,* e383–e417.

California Department of Education. (2013). Lead in paint. www.cde.ca.gov/ls/fa/hs/leadpaint.asp. Accessed 12 March 2012.

Carson, R. (1962). *Silent spring.* Boston: Houghton Mifflin.

Clark, B., & Jorgenson, A. (2012). The treadmill of destruction and the environmental impacts of militaries. *Sociology Compass, 6,* 557–569.

Clark, B., Jorgenson, A., & Kentor, J. (2010). Militarization and energy consumption. *International Journal of Sociology, 40,* 23–43.

Clark, B., Jorgenson, A., & Auerbach, D. (2012). Up in smoke: The human ecology and political economy of coal consumption. *Organization & Environment, 25,* 452–469.

Coe, M., Costa, M., & Soares-Filho, B. (2009). The influence of historical and potential future deforestation on the stream flow of the Amazon river-land surface processes and atmospheric feedbacks. *Journal of Hydrology, 369,* 165–174.

Colborn, T., Dumanoski, D., & Myers, J. (1996). *Our stolen future.* New York: Plume.

Cramer, W., Bondeau, A., Schaphoff, S., Lucht, W., Smith, B., & Sitch, S. (2004). Tropical forests and the global carbon cycle: Impacts of atmospheric carbon dioxide, climate change and rate of

deforestation. Philosophical transactions of the Royal Society of London. *Series B: Biological Sciences, 359,* 331–343.
de Marchi, S., & Hamilton, J. (2006). Assessing the accuracy of self-reported data: An Evaluation of the toxics release inventory. *Journal of Risk and Uncertainty, 32,* 57–76.
Del Olmo, R. (1998). The ecological impact of illicit drug cultivation and crop eradication programs in Latin America. *Theoretical Criminology, 2,* 269–278.
Ehrhardt-Martinez, K., Crenshaw, E., & Jenkins, C. (2002). Deforestation and the environmental kuznets curve: A cross-national investigation of intervening mechanisms. *Social Science Quarterly, 83,* 226–243.
Ellefsen, R., Sollund, R., & Larsen, G. (Eds.). (2012). *Eco-global crimes contemporary problems and future challenges*. Farnham: Ashgate.
Environmental Protection Agency (EPA). (2013a). Lead in drinking water. http://water.epa.gov/drink/info/lead/index.cfm. Accessed 15 March 2012.
Environmental Protection Agency (EPA). (2013b). National priorities list. www.epa.gov/superfund/sites/npl/index.htm. Accessed 15 March 2012.
Fall, S., Niyogi, D., Gluhovsky, A., Pielke, R., Kalnay, E., & Rochon, G. (2010). Impacts of land use land cover on temperature trends over the continental United States: Assessment using the North American regional reanalysis. *International Journal of Climatology, 30,* 1980–1993.
Foster, J. (2000). *Marx's ecology: Materialism and nature*. New York: New York University Press.
Foster, J. (2012). The planetary rift and the new human exemptionalism: A political-economic critique of ecological modernization theory. *Organization & Environment, 25,* 211–237.
Foster, J., & Burkett, P. (2008). Classical marxism and the second law of thermodynamics: Marx/Engels, the heat death of the universe hypothesis, and the origins of ecological economics. *Organization & Environment, 21,* 3–37.
Foster, J., Clark, B., & York, R. (2010). *The ecological rift: Capitalism's war on the earth.* New York: Monthly Review Press.
Gereffi, G. (1999). A commodity chains framework for analyzing global industries, Institute of Development Studies, 1999, Background Notes for Workshop on Spreading the Gains from Globalisation. http://eco.ieu.edu.tr/wp-content/Gereffi_CommodityChains99.pdf. Accessed 1 April 2012.
Gilbert, S., & Weiss, B. (2006). A rationale for lowering the blood lead action level from 10-2 pg/dL. *NeuroToxicology, 27,* 693–701.
Gould, K., Pellow, D., & Schnaiberg, A. (2008). *The treadmill of production: Injustice & unsustainability in the global economy*. Boulder: Paradigm.
Griffith, J., Stehman, S., & Loveland, T. (2003). Landscape trends in mid-atlantic and southeastern United States ecoregions. *Environmental Management, 32,* 572–588.
Hall, M. (2013). *Victims of environmental harm: Rights, recognition and redress under national and international law*. London: Routledge.
Harding, J. (2003). Historic deforestation and the fate of endemic invertebrate species in streams. *New Zealand Journal of Marine and Freshwater Research, 37,* 333–345.
Harvey, C., Medina, A., Sanchez, D., VmcHEz, S., Hernandez, B., Saenz, J., Maes, J., Casanoves, F., & Sinclair, F. (2006). Patterns of animal diversity in different forms of tree cover in agricultural landscapes. *Ecological Applications, 16,* 1986–1999.
Hillyard, P., & Tombs, S. (2005). Beyond Criminology? In P. Hillyard, C. Pantazis, S. Tombs, D. Gordon, & D. Dorling (Eds.), *Criminal obsessions: Why harm matters more than crime*. Sheffield: Crime and Society Foundation.
Hillyard, P., & Tombs, S. (2007). From crime to social harm. *Crime, Law and Social Change, 48,* 9–25.
Hubbell, S., He, F., Condit, R., Borda-de-Agua, L., Kellner, J., & ter Steege, H. (2008). How many tree species are there in the Amazon and how many of them will go extinct? *Proceedings of the National Academy of Sciences, 105,* 11498–11504.
Humphreys, D. (2003). Life protective or carcinogenic challenge? Global forests governance under advanced capitalism. *Global Environmental Politics, 3,* 40–55.
Jarrell, M., Lynch, M., & Stretesky, P. (2013). Green criminology and green victimization. In B. Arrigo & H. Bersot (Eds.), *The routledge handbook of international crime and justice studies.* London: Routledge.

Jeffery, C. (1977). *Crime prevention through environmental design*. Thousand Oaks: Sage.
Jorgenson, A. (2010). World-economic integration, supply depots, and environmental degradation: A study of ecologically unequal exchange, foreign investment dependence, and deforestation in less developed countries. *Critical Sociology, 36,* 453–477.
Jorgenson, A., & Clark, B. (2011). Societies consuming nature: A panel study of the ecological footprints of nations, 1960–2003. *Social Science Research, 40,* 226–244.
Jorgenson, A., Clark, B., & Kentor, J. (2010). Militarization and the environment: A panel study of carbon dioxide emissions and the ecological footprints of nations, 1970–2000. *Global Environmental Politics, 10,* 7–29.
Jorgenson, A., Clark, B., & Giedraitis, V. (2012). The temporal (in)stability of the carbon dioxide emissions/economic development relationship in central and eastern European nations. *Society & Natural Resources, 25,* 1182–1192.
Kanae, S., Oki, T., & Musiake, K. (2001). Impact of deforestation on regional precipitation over the Indochina peninsula. *Journal of Hydrometeorology, 2,* 51–70.
Kovel, J. (2007). *The enemy of nature: The end of capitalism or the end of the world?* London: Zed Books.
Landrigan, P., Schechter, C., Lipton, J., Fahs, M., & Schwartz, J. (2002). Environmental pollutants and disease in American children: Estimates of morbidity, mortality, and costs for lead poisoning, asthma, cancer, and developmental disabilities. *Environmental Health Perspectives, 110,* 721–728.
Lawton, R., Nair, U., Pielke, R., Sr., & Welch, R. (2001). Climatic impact of tropical lowland deforestation on nearby montane cloud forests. *Science, 294,* 584–587.
Li, Y., & Hewitt, C. (2008). The effect of trade between China and the UK on national and global carbon dioxide emissions. *Energy Policy, 36,* 1907–1914.
Linstad, H., Asbjornslett, B., & Stromman, A. (2011). Reductions in greenhouse gas emissions and cost by shipping at lower speeds. *Energy Policy, 39,* 3456–3464.
Long, M., Stretesky, P., Lynch, M., & Fenwick, E. (2012). Crime in the coal industry implications for green criminology and treadmill of production. *Organization & Environment, 25,* 328–346.
Lorion, C., & Kennedy, B. (2009). Riparian forest buffers mitigate the effects of deforestation on fish assemblages in tropical headwater streams. *Ecological Applications, 19,* 468–479.
Lovelock, J. (2000). *Gaia: A new look at life on Earth*. Oxford: Oxford Paperbacks.
Lovelock, J. (2007). *The revenge of gaia: Earth's climate crisis & the fate of humanity*. New York: Basic Books.
Lowe, A., Boshier, D., Ward, M., Bacles, C., & Navarro, C. (2005). Genetic resource impacts of habitat loss and degradation: Reconciling empirical evidence and predicted theory for neotropical trees. *Heredity, 95,* 255–273.
Lynch, M. (2013). Reflection on green criminology and its boundaries: comparing environmental and criminal victimization and considering crime from an eco-city perspective. In N. South & A. Brisman (Eds.), *The routledge handbook of international crime and justice studies*. London: Routledge.
Lynch, M., & Michalowski, R. (2006). *Primer in radical criminology*. New York: Criminal Justice Press.
Lynch, M., & Stretesky, P. (2010). Global warming, global crime: A green criminological analysis. In R. White (Ed.), *Global environmental harm and the greening of criminology*. Devon: Willan Publishing.
Lynch, M., McGurrin, D., & Fenwick, M. (2004). Disappearing act: The representation of corporate crime research in criminology journals and textbooks. *Journal of Criminal Justice, 32,* 389–398.
Lynch, M., Burns, R., & Stretesky, P. (2010). Global warming as a state-corporate crime: The politicalization of global warming during the Bush administration. *Crime, Law and Social Change, 54,* 213–239.
Magdoff, F., & Foster, J. (2011). *What every environmentalist needs to know about capitalism*. New York: Monthly Review Press.

Mandel, E. (1990). Karl Marx. In J. Eatwell, M. Milgate, & P. Newman (Eds.), *Marxian economics* (pp. 1–38). London: MacMillan.
Markowitz, G., & Rosner, D. (2003). Politicizing science: The case of the Bush administration's influence on the lead advisory panel at the Centers for Disease Control. *Journal of Public Health Policy, 24,* 105–129.
Marx, K. (1974). *Capital* (Vol. 1). Geneva: International Publishers.
Matthews, H., Hendrickson, C., & Horvath, A. (2001). External costs of air emissions from transportation. *Journal of Infrastructure Systems, 7,* 13–17.
Mbatu, R. (2010). Deforestation in the Buea-Limbe and Bertoua regions in southern Cameroon (1984–2000): Modernization, world-systems, and neo-malthusian outlook. *GeoJournal, 75,* 443–458.
McCollister, K., French, M., & Fang, H. (2010). The cost of crime to society: New crime-specific estimates for policy and program evaluation. *Drug and Alcohol Dependency, 108,* 98–109.
Millerd, F. (2011). The potential impact of climate change on great lakes international shipping. *Climatic Change, 104,* 629–652.
Needleman, H. (1997). Clamped in a straitjacket: The insertion of lead into gasoline. *Environmental Research, 74,* 95–103.
Negri, A., Adler, R., Xu, L., & Surratt, J. (2004). The impact of Amazonian deforestation on dry season rainfall. *Journal of Climate, 17,* 1306–1319.
Nicholls, R., Hoozemans, F., & Marchand, M. (1999). Increasing flood risk and wetland losses due to global sea-level rise: Regional and global analyses. *Global Environmental Change, 9,* S69–S87.
Nobre, C., Sellers, P., & Shukla, J. (1991). Amazonian deforestation and regional climate change. *Journal of Climate, 4,* 957–988.
Nurse, A. (2013). *Animal harm: Perspectives on why people harm and kill animals*. Farnham: Ashgate.
O'Connor, J. (1973). *The fiscal crisis of the state*. London: St Martin's Press.
Quinney, R. (1970). *The social reality of crime*. Piscataway: Transaction Publishers.
Ray, D., Nair, U., Lawton, R., Welch, R., & Pielke, R. (2006). Impact of land use on Costa Rican tropical montane cloud forests: Sensitivity of orographic cloud formation to deforestation in the plains. *Journal of Geophysical Research, Atmospheres, 111,* 1–16.
Reece, E. (2006). *Lost mountain: A year in the vanishing wilderness*. New York: Riverhead Books.
Rosner, D., & Markowitz, G. (1985). A "gift of God"? The public health controversy over leaded gasoline during the 1920s. *American Journal of Public Health, 75,* 344–352.
Schnaiberg, A. (1980). *The environment: From surplus to scarcity*. New York: Oxford University Press.
Scott, A. (1986). Industrialization and urbanization: A geographical agenda. *Annals of the Association of American Geographers, 76,* 25–37.
Sebbenn, A., Degen, B., Azevedo, V., Silva, M., Lacerda, A., Ciampi, A., Kanashiro, M., Carneiro, F., Thompson, I., & Loveless, M. (2008). Modelling the long-term impacts of selective logging on genetic diversity and demographic structure of four tropical tree species in the Amazon forest. *Forest Ecology and Management, 254,* 335–349.
Sherman, L., Gartin, P., & Buerger, M. (1989). Hot spots of predatory crime: Routine activities and the criminology of place. *Criminology, 27,* 27–56.
South, N. (1998). A green field for criminology? A proposal for a perspective. *Theoretical Criminology, 2,* 211–233.
Steingraber, S. (1997). *Living downstream*. New York: Random House.
Stretesky, P., & Lynch, M. (2002). Environmental hazards and school segregation in Hillsborough County, Florida, 1987–1999. *The Sociological Quarterly, 43,* 553–573.
Stretesky, P., & Lynch, M. (2009). A cross-national study of the association between per capita carbon dioxide emissions and exports to the United States. *Social Science Research, 38,* 239–250.
Stretesky, P., & Lynch, M. (2011). Coal strip mining, mountain top removal and the distribution of environmental violations across the United States, 2002-2008. *Landscape Research, 36,* 209–230.

Stretesky, P., Long, M., & Lynch, M. (2013a). Does environmental enforcement slow the treadmill of production? The relationship between large money penalties and toxic releases within offending corporations. *Journal of Crime and Justice, 36,* 235–249.
Stretesky, P., Long, M., & Lynch, M. (2013b). *The treadmill of crime: Political economy and green criminology*. London: Routledge.
Sweeney, B., Bott, T., Jackson, J., Kaplan, L., Newbold, J., Standley, L., Hession, W., & Horwitz, R. (2004). Riparian deforestation, stream narrowing, and loss of stream ecosystem services. *Proceedings of the National Academy of Sciences of the United States of America, 101,* 14132–14137.
Van Rompaey, A., Covers, G., & Puttemans, C. (2002). Modelling land use changes and their impact on soil erosion and sediment supply to rivers. *Earth Surface Processes and Landforms, 27,* 481–494.
Walters, R. (2010). Toxic atmospheres, air pollution, trade and the politics of regulation. *Critical Criminology, 18,* 307–323.
Wedeen, R. (2009). The lead industry and public health. *International Journal of Occupational and Environmental Health, 15,* 195–201.
Westerhuis, D., Walters, R., & Wyatt, T. (Eds.) (2013). *Emerging issues in green criminology: Exploring power, justice and harm*. Basingstoke: Palgrave Macmillan.
White, R. (2011). *Transnational environmental crime: Toward an eco-global criminology*. London: Routledge.
White, R. (Ed.) (2012). *Climate change from a criminological perspective*. Berlin: Springer.
Williams, C. (1996a). An environmental victimology. *Social Justice, 23,* 16–40.
Williams, C. (1996b). Environmental victimization and violence. *Aggression and Violent Behavior, 1,* 191–204.
York, R., Ergas, C., Rose, E., & Dietz, T. (2011). It's a material world: Trends in material extraction in China, Indonesia and Japan. *Nature and Culture, 2,* 103–122.
Zedler, J., & Kercher, S. (2005). Wetland resources: Status, trends, ecosystem services, and restorability. *Annual Review of Environmental Resources, 30,* 39–74.
Zhang, H., Henderson-Sellers, A., & McGuffie, K. (1996). Impacts of tropical deforestation. Part I: Process analysis of local climatic change. *Journal of Climate, 9,* 1497–1517.
Zheng, F., He, X., Gao, X., Zhang, C., & Tang, K. (2005). Effects of erosion patterns on nutrient loss following deforestation on the Loess Plateau of China. *Agriculture, Ecosystems & Environment, 108,* 85–97.
Zilney, L., McGurrin, D., & Zahran, S. (2006). Environmental justice and the role of criminology: An analytic review of 33 years of environmental justice research. *Criminal Justice Review, 31,* 47–62.
Zou, X., Zucca, C., Waide, R., & McDowell, W. (1995). Long-term influence of deforestation on tree species composition and litter dynamics of a tropical rain forest in Puerto Rico. *Forest Ecology and Management, 78,* 147–157.
Zullig, K., & Hendryx, M. (2010). A comparative analysis of health-related quality of life for residents of US counties with and without coal mining. *Public Health Reports, 125,* 548–555.

Dr. Michael J. Lynch is a professor of criminology at the University of South Florida in Tampa, Florida.

Dr. Michael A. Long is a senior lecturer in criminology at the Northumbria University in Newcastle, UK.

Dr. Kimberly L. Barrett is an assistant professor at the Eastern Michigan University in Ypsilanti, Michigan.

Dr. Paul B. Stretesky is a professor of criminology at the Northumbria University in Newcastle, UK.

Part III
Corporations and Brown Crime

Chapter 8
Accidents with Dangerous Substances in the Dutch Chemical Industry

Marieke Kluin

Introduction

Incidents such as Bhopal, Texas City Refinery and Deepwater Horizon show us that the consequences of work done by inspectors of chemical corporations is of extreme importance and affects daily life. It is in the public interest to get more insight into how these street-level bureaucrats arrive at their conclusions. How do they make these decisions? What are the factors relevant to the practical definition of compliance? How is compliance defined by these inspectors in their routine work? The focus of this study is on the implementation of regulations through enforcement by field-level inspectors from three different inspectorates: the Environmental Protection Agency, the Occupational Safety and Health Inspection Agency and the Fire Department. Hutter calls them gatekeepers of the regulatory process, since they are in the front line and determine how regulation is translated into action (1988). These individuals are responsible for the execution of policy and, according to Pautz (2010), for its success and failure.

Seveso inspections in the Netherlands are the objects of this research. These inspections have been performed for 10 years under the auspices of the Seveso directive on major hazards within the European Union. The inspections are carried out annually by inspectors from the Environmental Protection Agency, the Occupational Safety and Health Inspection Agency and the Fire Department in the Netherlands. Despite these inspections there are still many mistakes, accidents and even deliberate incidents in the chemical industry. Accidents can be organizationally or systematically created and could be viewed as crimes. This does not mean

This chapter originally appeared as a chapter in *Emerging Issues in Green Criminology: Power, Justice and Harm* edited by Walters et al. (2013).

M. Kluin (✉)
Leiden University, Leiden, The Netherlands
e-mail: m.h.a.kluin@law.leidenuniv.nl

T. Wyatt (ed.), *Hazardous Waste and Pollution*, DOI 10.1007/978-3-319-18081-6_8 that each and every industrial accident is the result of organizational crime (Pearce and Tombs 1998, p. 152). The focus of the current research is devoted to violations of environmental regulations by chemical corporations. One does not immediately associate this subject with criminals. Violations such as these do not fit within the normal range of offences according to the definition of crime in normal literature, press and crime statistics. Recent publications on the regulation of environmental offences show that these offences have typical characteristics and structures of enforcement (Kagan et al. 2011; Simpson and Rorie 2011).

There is little specific study on the role of field-level inspectors in the implementation of environmental regulations such as the one that Pautz (2010) did. This research begins to respond to this gap by focusing on a small subsection of field-level inspectors in environmental policy, environmental inspectors in the Netherlands. It is important to note that this study is broader than the activities of an isolated group of enforcement officials; it studies a complex social interactive process in which environmental health legislation is implemented. It is a detailed case study on how the Netherlands enforce specific European legislation with a focus on its implementation by field-level inspectors and compliance by regulatees.

The main purpose of this explorative part of the whole research is to obtain an understanding of environmental inspectors and their work, to collect more specific data about their working methods and to establish whether participant observation serve as a suitable method of investigation. Understanding these interactions will provide insights into how regulators approach their interactions with regulatees and how they ensure environmental compliance. McCaffrey et al. (2007, p. 308) note that 'understanding how relationships between regulatory organizations and firms affects compliance … is essential to understanding regulator implementation'.

The structure of this chapter is as follows: First, I discuss the theoretical background, the relevant literature on organizational crime, regulatory enforcement and the Seveso II directive. Second, I describe the inspectorate and the practice of environmental inspectors. Before presenting the results of participant observation at four Seveso corporations, I explain the research method and research setting of this case study. Then I describe the findings of the document analysis and participant observation of four Seveso corporations. In the concluding section of this chapter, the main findings are summarized.

Theoretical Background

Relevant Literature Organizational Crime

Organizational crime is one of the most difficult areas to study within criminology. This is due to its invisibility and complexity and it makes it even harder to investigate than common crime. Only a minority of offences is discovered and officially recorded. When one starts to investigate crime, most of the time the data is derived from official statistics, crime surveys and court records. These data do not include organizational crime, only conventional crime statistics, since not many organizational crimes are dealt with by the police. There is a long tradition within research on organizational crime that has its focus on industry case studies, for example: electricity (Geis 1967), pharmaceutical industries and coal mining industry (Braithwaite 1984, 1985), Bhopal (Pearce and Tombs 1998, 2012), criminogenic industries: oil, pharmaceutical and automobile industry according to Clinard and Yeager (1980), textile finishing and waste processing industry (Huisman 2001). However, the present study is also a study of particular forms of crime. There have been fewer attempts to focus upon particular forms of crime in any specific industry and in the view of Pearce and Tombs (1998) such work is especially useful of the understanding of the contours of organizational crime. A rare and therefore important example of this kind of research is their own research on corporate crime in the chemical industry. Another important inspiration for this research is Braithwaite's (1985) study on the coal mining industry. He studied the organizational differences between coal mining corporations that often violated occupational safety regulations and those that showed great willingness to comply with the regulations. He tried to determine how better compliance and therefore improved safety can be achieved. Braithwaite did his study to gain knowledge of the background of the causes and he analysed the safety compliance systems of the five corporate leaders at that time in coal mine safety. Hale states similar questions for example 'if the regulated cannot be required by law to do no harm, what should they be required to do and what is the role of the regulators? Should they be seeking to punish or persuade?' (Hale et al. 2002, p. 3). Braithwaite concludes that disasters in mines could be reduced if there is more willingness to comply with the law. An outcome of his study was that these disasters were most of the time the result of organizational crime. His research shows that a combination of punitive and educative measures taken against offenders can have substantial effects in reducing injuries to miners. Braithwaite found that in the safest coal mine companies safety was an aspect that was carried out by all the departments of the mine. He concluded that the overall factor in the five safest companies was the approach of the companies to guarantee that all departments comply with safety regulations.

Relevant Literature Regulatory Enforcement

This research takes a closer look at environmental inspectors, which are examples of street-level bureaucrats. Lipsky (1980) introduced the term 'street-level bureaucrat'; it refers to a public agency employee who actually performs the actions that implement laws. Hutter (1988, p. 4), whose research was an inspiration as well, says that environmental inspectors are in a position to determine to some extent what constitutes an offence or problem, and moreover to select which cases may be deemed suitable for legal action. 'Regulatory law is often vague, involving broad legal standard and the exercise of discretion by officials' (Hutter 1997, p. 3). Field-level inspectors have a great deal of discretion. They need to deal with a wide range of rules and interpret them on real-life situations and in addition their work is performed in the field in order to make inspections at corporations on-site. Up until now much of the study of regulatory actors is focused on the nature of enforcement.

The literature available on regulatory enforcement mentions two types of enforcement styles. This refers to the behaviours of inspectors when interacting with regulated entities (May and Winter 2011). On one side of the spectrum an inspector may decide for a 'punitive', 'rule oriented' or 'strict' enforcement (Bardach and Kagan 1982; Shover et al. 1984) which emphasizes on a more punitive means of achieving compliance. The other end of the continuum has been labelled as 'compliance' (Hawkins 1984) or 'accommodative' (Hutter 1988, 1989) oriented approach which can be characterized by using more cooperative or conciliatory means. Regulators were thought to operate under one of these approaches; however, research demonstrates that they apply these enforcement styles in varying combinations depending on circumstances, settings and policy areas (Hutter 1988, 1989; Mascini and van Wijk 2009a; May and Winter 2000, 2011). These insights are valuable, however, the focus of attention is on regulatory actors in isolation and it is unable to describe how regulators and regulatees interact. Interaction of environmental inspectors and regulatees fits within broader discussions of regulation.

According to Hutter (1997) and May and Wood (2003), insight of the regulatory arena and especially the important role of inspectors at the front lines of regulatory activity is missing. The present study contributes to the understanding of how those street-level bureaucrats of regulatory enforcement help to bring about compliance with regulations. May and Wood (2003, p. 118) state that:

> within broad parameters established by agency guidelines and routines, inspectors typically have a good deal of autonomy and discretion in how they go about regulation. They have often discretion over the thoroughness with which they conduct inspections over and over the manner in which they interact with regulatees. Street-level bureaucrats often make choices on the spot about the best mode of action to produce a desired set of behaviours from a target individual or group.

In line of the research done by Hutter (1988, 1997) and May and Wood (2003), the focus here is on interactions between field-level inspectors as street-level bureaucrats for environmental regulatory policy and regulatees.

Before taking a closer look at the inspectorate and the field-level inspectors of the Environmental Protection Agency, this chapter continues with relevant literature on enforcement of the European Union Directive, Seveso II. It is the legal framework in which the inspections take place.

Enforcement of the Seveso II Act in the Netherlands (Council Directive 96/82/EC)

In 1982, the European Union issued chemical industrial safety regulations as a reaction to chemical disasters in England (Flixborough 1974) and Italy (Seveso 1976). In Flixborough an explosion and a fire led to 28 fatalities and the destruction of the plant, and in the case of Seveso a vapour cloud containing dioxins escaped from a chemical plant. Member States of the European Union negotiated 3 years before the Seveso directive was adopted (Versluis 2003). This directive imposed stringent regulations on the industry. The main purpose of the Seveso regulations is to prevent and mitigate the effects of major accidents involving dangerous substances (Article 1, Seveso II directive). It regulates consequences of major accidents for the environment and on the other hand it also regulates the protection of employees of a corporation and also people outside the establishments. In the beginning of Seveso, the focus was on the technical part of safety at corporations. Later on it became clear that the cause of many industrial accidents were organizational aspects of safety, such as procedures and communication, especially since this directive did not prevent all accidents with dangerous substances like Bhopal (1984) and Basel (1986). These accidents outside the European Union led to two amendments of the directive in 1987[1] and 1988[2]. The first amendment was especially meant to prevent differing interpretations and clarified some entries and threshold levels. After the Basel accident which caused a major environmental pollution of the Rhine, the second amendment focused on the inclusion of isolated storage of dangerous substances.

Since the amendments were considered insufficient to guarantee the aim of preventing major accidents a complete revision of the directive was recommended by the European Commission and the European Parliament, and after long negotiations the second Seveso directive was accepted in 1996. It was updated in 1999, amended again in 2005 and is now referred to as the Seveso II directive. A change of focus on more general management systems was one of the main reasons and this change was too essential to regulate in an amendment according to Versluis (2003). Seveso II changed other elements as well compared to the first directive:

- The scope was broadened and simplified
- Seveso II focuses on entire establishments (Article 3, Seveso II directive) instead of individual installations
- Addition of land-use planning
- Extension of the article on inspections and member states are obligated to enforce the directive (Article 18, Seveso II directive)

It required operators of major hazard facilities to show the regulator that they had identified, assessed and controlled the hazards that are present in their facilities. The setting for this research is the regulation of chemical corporations in the Netherlands; it is called Besluit Risico Zware Ongevallen (BRZO)'99 (Besluit 1999), the Dutch implementation of the Seveso II directive. It integrates legislation in the field of occupational safety, external safety and disaster management within a legal framework and is an example of a coordination of laws. It represents the policy areas 'environment', 'occupational safety and health' and 'emergency planning'.

Corporations fall under the scope of Seveso II when the amounts of the permitted quantities of hazardous substances are above a certain level. There are two

[1] Directive 87/216/EEC of March 1987. OJ L 085, 28/03/1987, pp. 0036–0039.

[2] Directive 88/610/EEC of November 1988. OJ L 336, 07/12/1988, pp. 0014–0018.

thresholds, lower tier and upper tier. For each threshold, a limiting amount of dangerous substances is defined. Its objective is to prevent and control major accidents involving dangerous substances. A corporation has his own responsibility for organizing and securing the safety of its own activities. It is required to have a major accident prevention policy and a safety management system. If the dangerous substances exceed an upper-tier threshold, the corporation needs to provide a safety report to the authorities as well.

The actual enforcement of the Seveso II directive is a joint activity in the Netherlands and a Seveso II inspection team exists that consist of three inspectors of different agencies: the Environmental Protection Agency, the Occupational Safety and Health Inspection Agency and the Fire Department. During the combined inspection, inspectors with different backgrounds will look at the same corporation with various perspectives. A joint inspection is essential, because each inspector will have his own background, knowledge and 'looking glass' with his or her mandate and checks if the corporation complies with the rules. In the Netherlands, there is a shared uniform national inspection method that enables to run a Seveso II inspection. A joint inspection is also important for corporations: it results in fewer separate inspections.

At the end of a Seveso II inspection, inspectors will have an objective and reasoned opinion of the policy implemented by the corporation on major accidents and the safety management system. In short, the inspection team checks if a corporation:

- Has taken all measures necessary to prevent major accidents
- Has provided appropriate means for limiting the consequences
- The public has been informed
- Has data in the Safety Report that adequately reflects the real conditions:
  - Demonstrating that a major-accident prevention policy and a safety management system for implementing it have been put into effect
  - Demonstrating that major-accident hazards have been identified and that the necessary measures have been taken to prevent such accidents and to limit their consequences for man and the environment
  - Demonstrating that adequate safety and reliability have been incorporated into the design, construction, operation and maintenance of any installation, storage facility, equipment and infrastructure connected with its operation which are linked to major-accident hazards inside the corporation
  - Demonstrating that internal emergency plans have been drawn up and supplying information to enable the external plan to be implemented in order to take the necessary measures in the event of a major accident;
  - Providing sufficient information to the competent authorities to enable decisions to be made in terms of the siting of new activities or developments around existing corporations (Art. 9 conjunction Art. 18, Seveso II directive)

My present case study only focuses on the safety report requirements (Article 9) and inspection (Article 18). Other obligations that fall under the Seveso II directive—such as land-use planning legislation—are not included in this description of Seveso II enforcement. This chapter will take a closer look at only one of the agencies that is doing the Seveso inspections and the inspectors, the Environmental Protection Agency.

The Inspectorate: DCMR Environmental Protection Agency

The present research involves the most densely populated area of the Netherlands; it is called Rotterdam-Rijnmond. The DCMR Environmental Protection Agency is the regional environmental agency of the local and regional authorities operating in Rijnmond, the larger ‘Port of Rotterdam’ area in the Netherlands. The DCMR was founded in 1972 in order to improve environmental protection in the Rotterdam-Rijnmond region. This region is a heavily industrialized area with refineries, waste incinerators, several waste dumping sites, many large chemical plants, metallurgy, food processing plants and about 19,500 smaller companies. Seveso inspections are applicable for about 140 corporations in the Rotterdam-Rijnmond area. The tasks of the DCMR include regulation of the industries, monitoring and assisting authorities on developing environmental policy.

The DCMR divides its tasks among different divisions: municipalities and small and medium-sized enterprises (SME), port of Rotterdam and industry and expertise centre. These divisions have subdivisions like ‘monitoring and enforcement industry’ or ‘permits port and waste’. This study was limited to the two subdivisions that are dealing with Seveso inspections, because they mainly inspect chemical corporations who fall under the scope of Seveso. These subdivisions should inspect roughly 140 Seveso corporations every year. These corporations must take whatever measures necessary to prevent major accidents, to reduce consequences and they need to have a high level of protection for mankind and the environment. They need to set up an internal emergency plan, a major accident prevention policy, for the implementation and provision of a safety management system. The safety management system consists of a fixed number of defined elements. The way in which these elements are implemented in a corporation depends on the hazards and risks which are present at the Seveso corporation. In the next section, the practice of the inspectorate for the Seveso II directive is discussed before describing the method of research.

The Practice of the Inspectorate for the Seveso II Directive

Environmental inspectors of the DCMR are responsible for checking the compliance of upper and lower tier establishments. Since, a Seveso inspection team consists of representatives of three enforcement agencies, the inspectors participate in more than one team. Environmental inspectors do not have a routine office job and divide their time between the office and outside visits. They are in charge of multiple Seveso corporations and therefore inspecting premises, checking that improvements have been effected and investigating incidents.

Seveso inspections in the Netherlands are being performed annually at corporations during the time of my research. Seveso corporations receive a letter in advance announcing these inspections but for follow-up inspections the corporations are not always informed in advance and could be visited by surprise. Inspections are being performed at Seveso corporations to report any violations of the Seveso II directive. Inspectors check if the information in the safety report (if the corporation has one) conforms to the actual practice of the corporation.

Seveso inspections take place according to standardized phases:

1. preparation individual inspector: inspectors individually prepare their focus, subjects and themes, for example emergency procedures

2. preparation inspection team: inspection plan and schedule
joint determination of inspection subjects, safety management elements, themes and inspection goal

Elements of a Safety Management System (SMS):
SMS A. Major accident prevention policy
SMS B. Employment and organization
SMS C. Identification and evaluation of major hazards
SMS D. Operation control
SMS E. Management of change
SMS F. Planning for emergencies
SMS G. Monitoring performance
SMS H. Audit and review

3. carry out inspection: kick off, walk all around the plant and corporation, interviews, document review and close out

4. conclusion inspection: close out and inspection report
Inspection findings are categorized as:

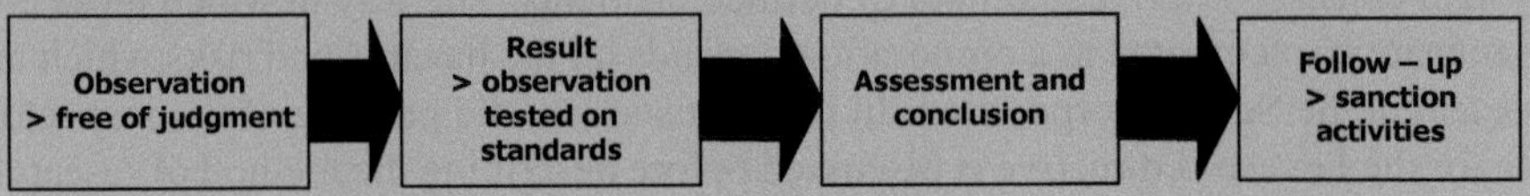

This overview is a simplified version of the whole inspection process (Werkwijzer BRZO 1999).

During such inspections, the above stated findings could lead to four possible conclusions. First, it is determined that the situation complies with the regulations. There will be no follow-up activities. Second, the outcome is sufficiently severe to lead to a violation. In this situation the corporation is explicitly responsible to carry out improvements. Third, it is determined that the situation does not comply with the regulations and it can be regarded as a violation of the law. Within the inspection team there will be a mutual agreement on who will proceed with sanctioning. The corporation will be notified of this intention to sanction and after a period it is again controlled by the inspectors. Fourth and last option, there is also a possibility that some findings individually are not sufficient enough to constitute an offence, but taken together they point to a structural deficit in an element of the safety management system and pose a violation. The agencies inspect together, but they sanction according their own legislation and currently it is not a joint activity. The Environmental Protection Agency sanctions according to the Environmental Management Act. Before discussing the data collection, the method of this study is described.

Method

Participant observations were made during the annual Seveso inspections and the follow-up at four chemical corporations in the Netherlands. These inspections took place between November 2010 and December 2011. The daily activities in these inspections are normally hidden for the general public and we still know very little on how inspectors establish compliance (Hutter 1997). The observed inspections are the yearly inspections as required by the Seveso II directive (Article 18, Council Directive 96/82/EC).

Participant observation is done to gather data on the social interaction between environmental inspectors and chemical corporations. Confidentiality is paramount since corporations that were inspected may have to deal with enforcement procedures. Since the objective of this research is to find out how inspectors make decisions, interpret the law and apply the law the 'atmosphere' in the interview room was important. The observations that were performed took place in areas where confidentiality was required to obtain information about, among other things, standard operating procedures and incidents. I chose the research role of observer as participant since I had to reveal my identity as an observer. In this approach, it is possible that the presence of the researcher is revealed by his informants to the setting, but does not really participate in the setting itself. The subjects under study are fully aware of the participation of the researcher. I was making notes but I did not participate in the inspection activities itself.

Observations were made of the interactions between the inspectors and regulatees during the daily routine of inspectors. Full notes were made while accompanying inspectors on their visits. These notes contained explicit references to participants, interactions, routines, rituals, temporal elements, interpretations, roles, working environment and behaviours. Data were not categorized while this exploratory phase of field work was still in progress. According to Hutter (1988, p. 210), one of the dangers of pre-classification is that you can blind yourself to new information so that it does not fall into an existing structure. The observed points of interest chosen in this research are not only based on literature, but also on an analysis of field notes after the phase of observing at four Seveso corporations. Although the list of points of interest will frame the analysis of the chemical corporations, I continued to record the observations broad, unfocused and general to have a good base for future lines of research (Hutter 1988).

The research section of participant observation is divided into two main categories:

A. *Standardized phases of Seveso II inspections*: observations were made during the inspection process when inspectors were assessing a fixed number of points at each corporation (Mascini and van Wijk 2008, 2009a, b).
B. *Points of interest*: research indicates that the following points of interest are relevant when doing participant observations of enforcement practices of regulation by field-level inspectors (Hutter 1988, 1997; Mascini and van Wijk 2008, 2009a, b).

The points of interest in this research are:

- *Inspector's perception of their role in this inspection process*. There is still very little written about environmental inspectors and the nature of their work. Since they have an important role in this inspection process it is crucial to know what their perception of their role is. Besides that, the opinion of the inspectors of the regulated might influence their enforcement.
- *Attitudes towards Seveso legislation*. Since this research studies on how the Netherlands enforce specific European legislation it is important to know what attitude inspectors and regulatees have towards this legislation. Versluis (2003) concludes that a positive opinion of inspectors of the legislation stimulates enforcement to actually take place and if they consider it important inspectors will put more time into enforcement. A negative opinion is likely to result in disinterest or hardly any enforcement. The opinion of the corporation might influence their compliance behaviour.
- *Proactive and reactive situations*. Offences are known by the Environmental Protection Agency in two ways: reported by a third party or discovered by an environmental inspector. In the document analysis and the participant observation component of this research both will be highlighted.
- *Working environment*. This topic not only refers to outside visits of inspectors while they are inspecting corporations but also to the working environment of the various inspectors at offices or at meetings with other inspection agencies.
- *Relationship with other agencies*. Since Seveso inspections are carried out by three different inspectorates: the Environmental Protection Agency, the Occupational Safety and Health Inspection Agency and the Fire Department, this was another central point in the observations.
- *Attitude towards violators*. Since participant observations are done in the setting of an inspection this research will describe the attitude of inspectors towards violators of the Seveso directive.
- *Type of offence*. A sanctioning practice of legislation is an interesting subject when studying enforcement. Seveso legislation is a European directive and therefore even more specific than general legislation. In the Netherlands three enforcement agencies inspect together, but as mentioned earlier, they sanction according their own legislation. What kind of environment offences are there and how do environment inspectors sanction them?
- *Attitude towards small and large corporations*. The Seveso II directive involves two thresholds, an upper and lower tier which is mentioned before. Both kinds of

corporations are studied in this explorative part of the research. Koolhaas (1990) observed that size of the regulated influence the style of an inspector. He states that large corporations are complex and inspectors are dependent on the good will of a corporation to co-operate. One of the questions that was posed by Versluis (2003) is investigated in this research, namely: To what extent do inspectors consider corporation size in their sanctioning practice?

- *Negotiations*. According to Manning (1988), compliance is 'the process of extended and endless negotiation'. Negotiations have been identified by numerous studies (Carson 1970; Cline 2010; Hawkins 1984; Hutter 1988; McAllister 2009; Shover et al. 1986) as a unique feature of regulatory enforcement. In the Netherlands negotiations could be even more of an issue since commission 'Oosting' concluded that inspectors should act more as enforcers and the 'gedoog-culture' of negotiations with regulatees should be abandoned immediately (2001). To what extend are negotiations between inspectors and Seveso corporations of influence on compliance?

Data Collection

The primary data for the explorative part of this research was collected between November 2010 and December 2011. The data consists of participant observations, interviews and document analysis. The research started in September 2010 with selecting a sample of chemical corporations within the 140 Seveso corporations of the two subdivisions. The objective of this selection was that the sample needed to be representative of the variety of work undertaken by Seveso II inspectors and of all the Seveso II corporations in this area. Since the intention was to observe whole inspection processes, this selection process has practical choices. A Seveso inspection starts with a preliminary meeting among the inspectors of the Environmental Protection Agency, the Occupational Safety and Health Inspection Agency and the Fire Department to determine the inspection subjects. Since this meeting forms the basis of the whole inspection process, I needed to observe these meetings too. A second choice was made on the basis of practicality. For the preparation, the Seveso II inspection team needs half a day, the inspection itself could have a duration of 7 days and after the inspection they needed half a day as well. I decided to observe one Seveso II inspection each month. On that basis, the sample exists of 15 Seveso II corporations and this chapter describes the results of 4 of these establishments; this is the explorative part of the complete research.

It took approximately 10 months to get permission of the inspection agencies and permission of the corporations, the inspection of which would be observed and to be present during the inspections. In addition to that, I needed to have personal protective equipment to enter the chemical corporations, a personal safety logbook, obligatory safety trainings and exams. After the training and exams I had more knowledge beforehand of the processes and did a small part of the safety training of employees of chemical establishments. Of course, I needed to know as well what to do when sulphide chemical would accidentally be released during an inspection. All corporations and inspectors co-operated with the research.

Data Analysis

Registered Offences and Enforcement Activities of Four Seveso Corporations

Table 8.1 is an impression of the corporations that are under study in this research. In the table are some key features outlined, such as the nature of the process and the number of employees. As Table 8.1 shows the processes of the corporations are different, as well as the age of the corporation and the number of employees. Since Seveso II has two thresholds it is a normal characterization of these laws and regulations. As mentioned earlier, this distinction is determined based on the permitted quantities of dangerous substances.

The database and archive of the Environmental Protection Agency is used for the document analysis. All the reports for the four corporations were selected that were available since the beginning of Seveso in the Netherlands (1999). Since knowledge on these two resources is limited I started checking every report. Throughout this process, I learned that environmental inspectors made a decision if an incident was an offence or not. Since this research only concentrates on violations of environmental regulations of chemical corporations I decided to look exclusively at offences in these resources. From some corporations I could find reports from offences and follow-up activities of notices, penalties and lawsuits starting from 2000, but not from all the corporations. I used all the available data on the offences recorded by the Environmental Protection Agency of the four corporations between 1999 and 2011 to get a complete overview of the environmental offences which is presented below in Fig. 8.1.

Figure 8.1 shows all the environmental offences the four corporations committed between 1999 and 2011. A total of 257 offences were committed in this timeframe. The immediate cause of these offences can be categorized as: complaints made by citizens or neighbouring corporations, the corporation itself (by submitting reports or making mentions of activities), the Environmental Protection Agency

Table 8.1 Overview of the four Seveso corporations

C01 process: filling and storage of propane, carbon dioxide and storage of industrial gases in cylinders age: less than 10 years number of employees: less than 50 Seveso: lower tier	C02 process: transportation, warehousing and handling containers with hazardous substances age: more than 40 years number of employees: less than 250 Seveso: lower tier
C03 process: producer of industrial gases of hydrogen and carbon monoxide age: more than 10 years number of employees: less than 50 Seveso: lower tier	C04 process: refinery age: more than 40 years number of employees: more than 250 Seveso: upper tier

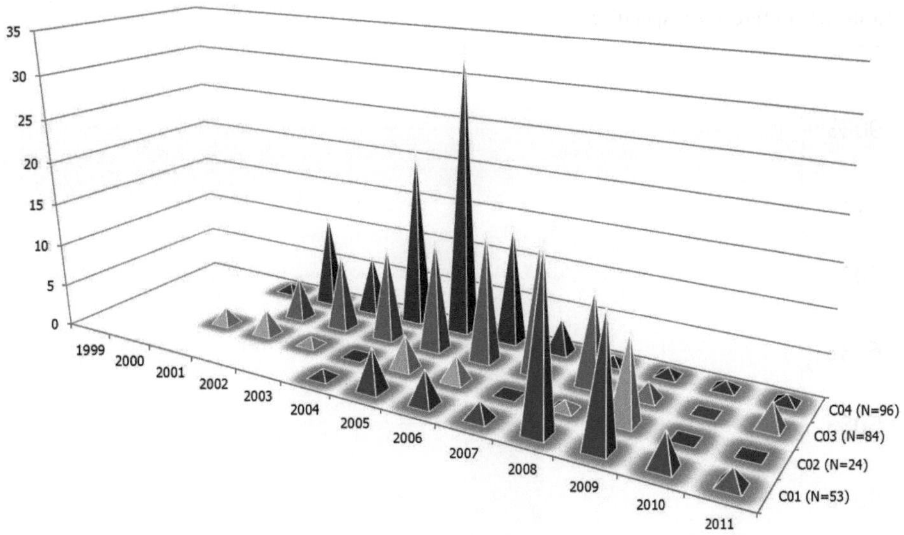

Fig. 8.1 Total environmental offences of four Seveso corporations

Table 8.2 Enforcement overview

	C01	C02	C03	C04
1999		1		
2000		2		
2001		2		
2002		2	1	1
2003		1	2	0
2004	0	0	4	5
2005	1	0	1	3
2006	2	2	2	3
2007	1	1	3	4
2008	6	2	1	4
2009	8	1	3	1
2010	1	0	0	2
2011	1	1	2	3
Total	*20*	*15*	*19*	*26*

(by doing inspections or following up reports by corporations) and other causes. The backgrounds to these violations differ, for example: inappropriate storage of various hazardous substances, incidents and violations from regular inspections on certain requirements or regulations. By taking a closer look at 2007 and 2010, it can be noted that the total number of offences is lower than in other years. This research needs to have more information to draw conclusions on this change in the number of offences (Tables 8.2 and 8.3).

Table 8.2 reflects the follow-up activities of the Environmental Protection Agency on these violations. A total of 31% of all the offences are enforced. An eye-

Table 8.3 Enforcement specifics

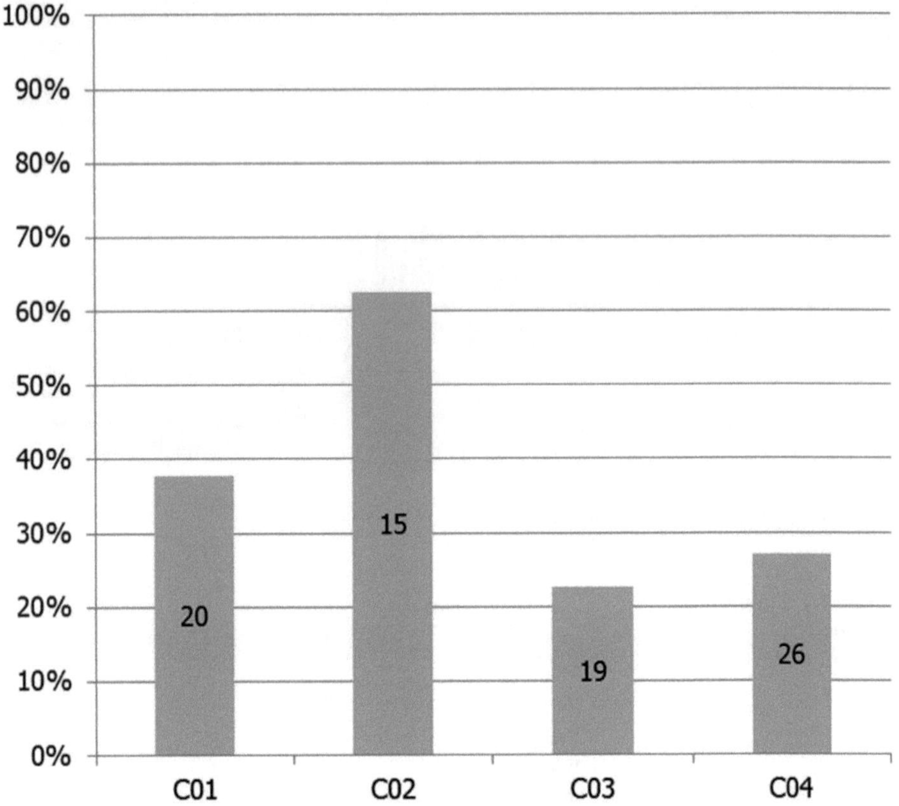

catching observation is the low percentage (27 %) of enforcement of corporation C04, which has the highest amount of offences ($N=96$). This low percentage can be explained by looking at the number of complaints; for this corporation the number of complaints about odour and noise is higher than for the other corporations. If a corporation was dealing with these issues and implementing a solution for the problems that cause the complaints, the Environmental Protection Agency did not enforce it. In this explorative part, I do not have a Seveso corporation of the same size to make a good comparison. This study needs to do more research to draw more robust conclusions on this.

When I studied the enforcement activities of the four corporations, I found out that not all of them are related to the environment offences in Fig. 8.1. This was an unsuspected finding. Table 8.3 shows that corporation C02 has a total of 24 offences and 63 % of them are enforced (and that are 15 enforcement actions). Of those, 15 enforcement actions, 9 enforcement activities are not related to the offences mentioned in Fig. 8.1. I studied all of the non-related enforcement activities and found out that there are three immediate causes for this. The causes are: reporting of the corporation itself, an incident, and the third during an inspection there was an observation and at the moment it was not a violation. In the end all of these causes lead to enforcement activities.

Since none of the above tables gives specific details on practice or details of the interactions between inspectors and regulatees participant observation is the other method of research that is used in this study. The next section, will describe the findings of participant observations at four Seveso corporations.

Participant Observation at Four Seveso Corporations

Observations were made during the daily routine of inspectors. I accompanied inspection teams at the preparation of an inspection, during an inspection and after a Seveso II inspection. A total of fourteen inspectors and eight inspections are observed at these four corporations. I observed 26 days in total for these eight inspections and saw seven different teams of inspectors and observed at one corporation three inspections, and at another corporation observed two inspections. Of these eight inspections, seven were normal Seveso II inspections and one was a follow-up. This follow-up inspection also fell under the scope of Seveso and was to determine if violations were ended. Of these fourteen inspectors, four of them were female, but this branch of the industry is dominated by males.

As mentioned earlier, the research section of participant observation is divided into two main categories: standardized phases of Seveso II inspections and points of interest. The next section will describe the participant observations of the Seveso corporations.

Participant Observation of Standardized Phases of Seveso II Inspections

First, a description of the observations of the standardized phases of Seveso II inspections is given.

Observations of Preparations of Inspections During the preparation of the inspection, inspectors decide which elements of the safety management system they are going to inspect. They decide that on the basis of what had already been inspected, so previous inspections, possible incidents and past reports of inspections. The whole inspection team decides which persons they would like to interview (by functions) during the inspection at the different elements and documents that needs to be sent by the corporation in advance. An average of six elements were discussed throughout one Seveso inspection. In the joint inspection system they have examples for control lists for each element, but inspectors make audit trails themselves as well (different observations between 2010–2011).

If inspectors have decided which safety management elements and other topics they are going to inspect, they discuss each subject in detail and apply it on the specific installations the corporation has. They discussed, for example, corporation C02 which scenario's they are going to inspect and how (observed January 2010).

Observations During Walks Around the Plants and Corporations Almost every inspection I observed started after the introduction with a walk around the plant. An inspection team is able to look into more details at small corporations such as C01 and C03. They can inspect details such as filling points, storage of acetylene and cylinders. During the walk around of the inspection at corporation C01 the inspection team notices that the employee who is working in the zone with risk of explosion carries a mobile phone that is turned on. He explains that he carries it to improve the corporation's contactability although it is not allowed (observed November 2010).

At large corporations this part is more difficult and inspection teams decide in advance what installation will be more thoroughly inspected although they see the whole site, but most of the time not by walking but cycling. The walk around at corporation C04 is a cycle tour around the plant. Employees of the corporation show us the highlights. We received a detailed description of each production unit and we could ask or discuss something in great detail. During this inspection we did another field check on the topic of temporary changes of the emergency equipment. Since they were randomly picked the corporation did not know this in advance and the inspection team has a more announced visit of these parts of the plant (observed January 2011).

These walks around the premises of a Seveso corporation give a good impression on the housekeeping and how these corporations work. Since corporations know that inspectors want to inspect in the field it is of course possible that they clean up in advance (different observations between 2010–2011)

Observations of Fixed Elements of Safety Management Systems (SMS) One corporation needed to show for example for safety management element B (employment and organization) if they gave instructions to their employees and what kind of training they gave. Of course, staff needs to have the right knowledge to perform their duties within the corporation. At one of the corporations, safety instructions were not clear and it was also not clear what each employee had done for his training. Inspector inspected this topic by asking corporation employees and a document analysis. This safety management element needs to be updated by the corporation since certain trainings have obligated certificates and if it is not there, the employee cannot perform his task (observed November 2010).

Some corporations have procedures that are applied worldwide and are difficult to bring in line with Dutch rules and regulations (observed December 2010).

Inspection teams generally first start with the procedure of safety management element D (operation control), to check if it is updated to the current situation of the corporation, if it is clear and in compliance with the Seveso II directive. Corporations are obligated to identify major hazards (SMS element C) and therefore they could do safety reviews like failure mode, effects and criticality analysis (FMECA) and the result of the reviews be certain measures. These measures should be in a good condition (a fire system that actually works for example) and there should be maintenance systems to maintain these measures. A question of an inspector could be: is there a maintenance management system in place (observed March 2011)?

Inspectors sometimes select an installation or a certain instrument and go through the whole safety management system to find out if this particular object is maintained, employees are certified and educated to work with it, if tests have been done regularly and adequate technical measures are in place (and working) to prevent major accidents (observed January 2011).

I observed different ways of inspecting the various safety management elements at these four corporations. An example is safety management element E, management of change, were the fire brigade department made a special theme around this element and inspect this at all Seveso inspections that year in the Rotterdam-Rijnmond area. They made a special audit trail to check how corporations deal with temporary fire extinguishers and fire systems. In this particular example, it involves means that are temporary shutdown, but are in a normal situation available. How do corporations communicate that within their corporation, what do they put in place and what are the consequences for operating the installations (observed March 2011)? In other inspections corporations needed to show the inspectors how they approach this and use it within their corporation. Larger corporations have whole systems for this, since changes could happen every day, for example changing of a pump, and more people needed to approve this change. Sometimes they first needed to do a safety review, before they decide to change their operation (observed January 2011)

A question an inspector could ask for safety management element G, monitoring performance is: Are there procedures for an ongoing systematic assessment of safety performance in relation to the objectives of the policy to prevent major accidents? The smaller corporations in this explorative part have difficulties in dealing with this element and both had violations. The larger ones have whole systems for it to keep track on their output, especially for safety performance indicators (different observations between 2010 and 2011).

Observations of Conclusions of Inspections After the inspection days at the corporation the inspection team discusses the results among themselves. They prepare a presentation on main findings and present the results of the inspection to the corporation. At corporation C01, the inspectors discusses the main findings just after the last inspection subject without the presence of the corporation. After that they present their main findings on the same day (observed November 2009). At all the other inspections the inspectors have a meeting on another day in advance of the presentation at the corporation. At those meetings they discuss the inspection results and make a presentation together. Then there is room for discussion on the inspection results and time to talk details thorough (different observations between 2010–2011).

After that the team of inspectors must draw its conclusion in an end report. In three cases of these eight inspections the inspectors needed to resort to enforcement activities because of violations of Seveso regulations. They sometimes made these decisions together and sometimes they do the reinspections together as well.

The offenses were on the following elements: organizational and personnel (B), identification and evaluation of major hazards (C), operational control (D), planning for emergencies (F), monitoring performance (G) (different observations between 2010–2011).

Participant Observation of Points of Interest

The second main category of the research component of participant observation is 'points of interest' which is described below.

Inspectors' Perception of their Role in this Inspection Process Hutter says that inspectors of law enforcement agencies have a variety of tactics and enforcement devices at their disposal, ranging from informal, conciliatory techniques to the formal and more coercive tools (1988). During the eight inspections I observed at the four corporations, I saw multiple styles, even among inspectors of the same inspection team.

For example at corporation C01, the inspection leader had a formal and more strict role. He inspected safety management element B, employment and organization, and after a few questions it was clear the corporation did not have any knowledge on these specific regulations. The inspectors had two roles at that inspection: they clearly explained what the corporation did wrong and how they could improve the mistakes and helped them complying with the rules, but they were also checking the 'old violations' and more behaving as 'police-officers'. In this specific situation the 'police-offer' role was the inspection leader of the inspection team. The violations of the corporation continued for a longer period (observed November 2009).

Attitudes Towards Legislation of Inspectors When I accompanied inspectors during their daily activities I drove with them to the corporations and we had conversations on this topic. Some inspectors say that the Seveso directive has too many rules and that they are not clear. Other inspectors have a different opinion and regard the Seveso directive as a flexible and therefore a good applicable directive since small and large corporations need to abide it.

Proactive and Reactive Situations Both situations appear in this research. During the walk around the corporation C04, the inspection team observed that pipes are hanging loose at an installation. During this outside visit the inspection team is accompanied by employees of the corporation and they notice it as well. After the walk around the inspectors and employees of the corporation discuss this observation. Corporations C04 took this observation seriously and immediately afterwards they send a team over there to check it in more detail. They find out that the pipes are in use and contain hydrogen. During this inspection period, the corporation reports that the problem is solved. The inspectors check it with a field check and it is correct. The offence does not show up in the final inspection report (observed January 2011).

There are two kinds of reactive situations: first, in response to complaints and secondly, in response to accidents and incidents. An example is: a police officer of the Harbour Police reports to the Environmental Protection Agency that corporation

C01 is doing business activities and he is alarmed by the fact that the corporation has dangerous goods. This report is the immediate cause for an environmental inspector to perform an inspection, since they were not aware of the existence of the corporation before that.

Working Environment Most of the time these corporations are located in an industrial area and not close by neighbourhoods. Some corporations have a barrier in front of their premises and at others you can easily walk in. All four corporations had some kind of a registration system for visitors (different observations between 2010 and 2011).

Before observing these inspections I needed to do an obligatory safety training which is combined with a safety passport. In this exploratory stage only one corporation asked me to show the passport to them. Of course this is probably related to the fact that I was always accompanied by inspectors, but it was strange that not all corporations demanded me to show it to them. Especially, since dangerous goods at all corporations were stored, produced or handled, and in full production while I was there (different observations between 2010 and 2011).

Relationship with Other Agencies Since Seveso inspections are a joint effort in the Netherlands, inspectors need to work together during the whole process of an inspection. Some inspection teams existed out of two inspectors, but most inspections I observed all three representatives of the agencies were involved. If there was any discussion among them, that always took place at the meeting after the inspection days and not in front of the corporation. These discussions were of course related to the findings during the inspection if they were offences or not (different observations between 2010 and 2011).

Negotiations Large corporations can and do use their power to manipulate the labelling process and minimize sanctions (Hutter 1997). The earlier mentioned offence of corporation C04 which did not end up in the inspection report is a good example of negotiations. On the other hand the inspector told me that the pipe contained a less dangerous product and he assessed the risk as low. This relates to the fact of practicability where inspectors are willing to accept anything if the necessary results are achieved.

Comparison of Document Analysis and Participant Observation

As already mentioned in this chapter, two data sources were used for this research: document analysis of registered offences and enforcement activities and participant observation of four Seveso corporations. This section shows the importance of both data sources in this research by describing two examples. In the document analysis there was a report of an offence of corporation C01 in 2005. These violations were noticed during an inspection; it involved a violation of using a mobile phone in an explosive area where it was not allowed. During the inspection of November 2010, inspectors interviewed an operator while he is working at an installation. They ask him if he carries a mobile phone and if it is currently working, and they find out that the mobile phone is turned on. This is a violation of the corporations' own rules and regulations and also a violation of the Environmental Management Act. The inspectors decided that the offence does not receive a follow-up or enforcement. I wonder whether, if they had known about the history of the offence in 2005, which makes the offence in 2010 a case of recidivism, if they would make the same decision by not to sanction the violation. Without the participant observation of the inspection of corporation C04, I could not have known that the corporation had a violation which is described in the part on proactive situations of this chapter, since the violation did not show up in the inspection report in the end. These two examples show that both data sources are valuable when studying the implementation of field-level inspectors, especially when you do a case study research on corporations. Since this chapter is the result of an explorative part of a complete research, we need to observe more inspections to see whether some of these findings are unique or repetitive. Document analysis only shows the end result of these proactive and reactive enforcement activities; participant observation gives a more complete view of what is really going in the field.

Conclusion

Seveso inspections in the Netherlands are the objects of this research. These inspections have been performed for 10 years under the auspices of the Seveso directive on major hazards within the European Union. The inspections are carried out annually by inspectors from the Environmental Protection Agency, the Occupational Safety and Health Inspection Agency and the Fire Department in the Netherlands. There is little specific study on the role of field-level inspectors in the implementation of environmental regulations. Our present research begins to respond to this gap by focusing on a small subsection of field-level inspectors in environmental policy, environmental inspectors in the Netherlands. The main purpose of this explorative part of the research is to obtain an understanding of environmental inspectors and their work, to collect more specific data about their working methods and to establish whether participant observation served as a suitable method of investigation. Understanding these interactions provided insights into how regulators approach their interactions with regulatees and how they ensure environmental compliance. The primary data for this research was collected between November 2010 and December 2011. The data consists of participant observations, interviews and document analysis. Participant observations were made during the annual Seveso inspections and the follow-up at four chemical corporations in the Netherlands. These inspections took place between November 2010 and December 2011. During these inspections of the premises, inspectors assess an impression of a part of a corporation. Since time is limited, they see a 'snapshot' of the corporation's condition, the abilities of the personnel and gained an impression of what was really going on. But this visible presence of inspectors is thought to hold a deterrent value, and it shows that inspectors enforce the law. Since this chapter is an explorative part of the complete research, we need to observe more inspections to see whether some of these findings are unique or repetitive. Document analysis only shows the end result of these proactive and reactive enforcement activities and participant observation gives a more complete view what is really going in the field.

Acknowledgments I would like to thank the Environmental Protection Agency and their inspectors for their generous cooperation in this study. I thank Ben Ale, Wim Huisman and Ellen Jagtman for their valuable comments on a previous version of this chapter.

References

Bardach, E., & Kagan, R. (1982). *Going by the book: The problem of regulatory unreasonableness*. Philadelphia: Temple University Press.

Besluit. (27 mei 1999). tot vaststelling van het Besluit risico's zware ongevallen 1999 en tot herziening van enkele andere besluiten in verband met de uitvoering van Richtlijn nr. 96/82/EG van de Raad van de Europese Unie van 9 december 1996 betreffende de beheersing van de gevaren van zware ongevallen waarbij gevaarlijke stoffen zijn betrokken, Staatsblad 1999, 234.

Braithwaite, J. (1984). *Corporate crime in the pharmaceutical industry*. London: Routledge & Kegan Paul.

Braithwaite, J. (1985). *To punish or persuade*. Albany: State University of New York Press.

Carson, W. G. (1970). Some sociological aspects of strict liability and the enforcement of factory legislation. *Modern Law Review, 33,* 396–412.

Clinard, M. B., & Yeager, P. C. (1980). *Corporate crime*. New York: Free Press.

Commissie-Oosting. (2001). *De vuurwerkramp*. Eindrapport van de Commissie Vuurwerkramp Enschede. Enschede, Den Haag.

Council Directive 96/82/EC on the control of major-accident hazards involving dangerous substances, OJ 1997 L 10/13.

Cline, K. (2010) Negotiating the superfund: are environmental protection agency regional officials willing to bargain with states? *The Social Science Journal, 47,* 106–120.

Directive 87/216/EEC. (1987). OJ L 085, 28/03/1987, pp. 0036–0039.

Directive 88/610/EEC. (1988). OJ L 336, 07/12/1988, pp. 0014–0018.

Geis, G. (1967). The heavy electric equipment antitrust case of 1961. In M. B. Clinard & R. Quinney (Ed.), *Criminal behaviour systems: A typology* (pp. 139–150). New York: Holt, Rinehart and Winston.

Hale, A., Hopkins, A., & Kirwan, B. (2002). *Changing regulation: Controlling risks in society*. Oxford: Pergamon.

Hawkins, K. (1984). *Environment and enforcement: Regulation and the social definition of pollution*. Oxford: Clarendon Press.

Huisman, W. (2001). *Tussen winst en moraal. Achtergronden van regelnaleving en regelovertreding door ondernemingen*. Den Haag: Boom Juridische Uitgevers.

Hutter, B. M. (1988). *The reasonable arm of the law*. Oxford: Clarendon Press.

Hutter, B.M. (1989). Variations in regulatory enforcement styles. Law & Policy, 11(2), 153–174. doi:10.1111/j.1467-9930.1989.tb00024.x.

Hutter, B. M. (1997). *Compliance: Regulation and environment*. Oxford: Clarendon Press.

Kagan, R.A., Gunningham, N., & Thornton, D. (2011). Fear, duty, and regulatory compliance: Lessons from three research projects. In C. Parker & V. L. Nielsen (Eds.), *Explaining compliance business responses to regulation* (pp. 37–58). Northampton: Edward Elgar.

Koolhaas, E. (1990). Milieu, handhavingsstrategieën en beleid. *Justitiële Verkenningen, 16,* 92–109.

LAT risicobeheersing bedrijven (2007) Werkwijzer BRZO. (1999). http://www.latrb.nl/instrumenten-0/werkwijzer-brzo-ii-0/. Accessed 8 Aug 2012.

Lipsky, M. (1980). *Street-level bureaucracy: Dilemmas of the individual in public services*. New York: Russell Sage Foundation.

Manning, P. K. (1988). [Review of Braithwaite, J.] To punish or persuade. *British Journal of Criminology, 28*(4), 559–561.

May, P. J., & Winter, S. (2000). Reconsidering styles of regulatory enforcement: Patterns in Danish agro-environmental inspection. *Law & Policy, 22,* 143–173.

May, P. J., & Winter, S. (2011). Regulatory enforcement styles and compliance. In C. Parker & V. L. Nielsen (Eds.), *Explaining compliance business responses to regulation* (pp. 222–244). Northampton: Edward Elgar.

May, P. J., & Wood, R. S. (2003). At the regulatory front lines: inspectors' enforcement styles and regulatory compliance. *Journal of Public Administration Research and Theory, 13*(2), 117–139.

Mascini, P., & van Wijk, E. (2008). 'Vis ruikt nou eenmaal zo'. Responsive regulation door de Voedsel en Waren Autoriteit. *Tijdschrift voor Criminologie, 50*(2), 114–129.

Mascini, P., & van Wijk, E. (2009a). *Responsive regulation bij de Voedsel en Waren Autoriteit*. Den Haag: Boom Juridische uitgevers.

Mascini, P., & van Wijk, E. (2009b). Responsive regulation at the Dutch food and consumer product safety authority: An empirical assessment of assumptions underlying the theory. *Regulation & Governance, 3,* 27–47.

McAllister, L. K. (2009). Dimensions of enforcement style: Factoring in regulatory autonomy and capacity. *Law & Policy, 32*(1), 61–78.

McCaffrey, D. P., Smith, A. E., & Martinez-Moyano, I. J. (2007). 'Then let's have a dialogue': Interdependence and negotiation in a cohesive regulatory system. *Journal of Public Administration Research and Theory, 17,* 307–334.

Pautz, M. C. (2010). Front-line regulators and their approach to environmental regulation in Southwest Ohio. *Review of Policy Research, 27*(6), 761–780.

Pearce, F., & Tombs, S. (1998). *Toxic capitalism: Corporate crime and the chemical industry*. Dartmouth: Ashgate.

Pearce, F. , & Tombs, S. (2012). *Bhopal: Flowers at the altar of profit and power*. Lincolnshire: CrimeTalk Books.

Shover, N., Lynxwiler, J., Groce, S., & Clelland, D. (1984). Regional variation in regulatory law enforcement: The surface mining control and reclamation act. In K. Hawkins, K. Thomas, & J. Thomas (Eds.), *Enforcing regulation* (pp. 121–146). Boston: Kluwer-Nijhoff.

Shover, N., Clelland, D., & Lynxwiler, J. P. (1986). *Enforcement or negotiation? Constructing a regulatory bureaucracy*. Albany: State University of New York Press.

Simpson, S.S., & Rorie, M. (2011). Motivating compliance: Economic and material motives for compliance. In C. Parker & V. L. Nielsen (Eds.), *Explaining compliance business responses to regulation* (pp. 59–77). Northampton: Edward Elgar.

Versluis, E. (2003). *Enforcement matters*. Utrecht: Utrecht University.

Dr. Marieke Kluin is a lecturer in Criminology at Leiden University in Leiden, the Netherlands.

Chapter 9
Cleaning Up Greenwash: A Critical Evaluation of the Activities of Oil Companies in the Niger

Angus Nurse

Introduction

Green criminology identifies that corporate environmental crimes 'often eclipse the scope and reach of the criminal law' (Sollund 2012, p. 3) as many environmentally harmful activities fall outside of the strict definition of crime as acts prohibited by the criminal law (Situ and Emmons 2000). Instead, they amount to regulatory non-compliance or technical breach outside the remit of mainstream policing agencies. Ideologically, this raises questions about how best to deal with non-compliance by ostensibly legal actors where a lack of regulatory control fails to encourage environmentally responsible behaviour (White and Heckenberg 2014). While many corporations embrace the concepts of social and environmental responsibility, numerous examples exist of corporations who claim to act in a sustainable and responsible way while at the same time causing considerable environmental damage (Doyon 2014; Nurse 2014). White (2012, p. 15) identifies that 'the systematic causal chains that underpin much environmental harm are located at the level of the global political economy—within which the transnational corporation stands as the central social force'. Globalisation's increased flows of people, products and profits allow multinational corporations (MNCs) to operate freely in developing countries where regulatory controls may be weaker than in the developed world and where public institutions (including government bureaucracy and the judiciary) may be weak and susceptible to economic pressure (Cox, in McGrew 1997, p. 51). Green criminology is frequently concerned with environmental harm, the adverse consequences of human behaviour upon the environment and non-human nature (Beirne and South 2007; Benton 1998; Lynch and Stretesky 2014). The global operations of MNCs can have significant negative consequences for the communities in which they operate and the wider environment yet are often legal given the relative lack of regulation for corporate practice in relation to environmental harm. Lawful activities

A. Nurse (✉)
School of Law, Middlesex University, The Boroughs, London NW4 4BT, UK
e-mail: a.nurse@mdx.ac.uk

T. Wyatt (ed.), *Hazardous Waste and Pollution,* DOI 10.1007/978-3-319-18081-6_9

such as oil extraction may result in harmful consequences for ecosystems which may fall outside the strict legalistic definition of crimes (Situ and Emmons 2000) where these are unintended or accidental consequences of a lawful activity. Arguably neoliberal markets encourage such activities by valuing profit and ignoring or minimising externalities, particularly the wide range of environmental harms that arise from ostensibly lawful activities which may result in pollution incidents such as oil spills (Lynch and Stretesky 2014, p. 8–10; Dybing 2012).

This chapter argues that responsibility for environmental damage is both a corporate and social responsibility, but is one for which international law mechanisms and self-regulation have proven to be inadequate. Examining the actions of oil companies in the Niger Delta, this chapter applies a green criminological perspective to corporate environmental responsibility (CER), arguing that environmental harms committed by legal actors have the potential to cause greater harm than criminal activities and thus should be the subject of regulation to ensure corporate responsibility for environmental damage. However, the chapter's application of a green criminological perspective also identifies that action taken outside of the criminal law to address such harms, particularly by way of civil claims, is potentially more effective in achieving redress than the limited punishment-based approach of the criminal law.

The Corporate Environmental Responsibility Landscape

White (2007) identifies inequality as a significant aspect of environmental justice, particularly the manner in which ethnic minority and indigenous communities suffer at the hands of western forces. Decades of oil exploration in the Niger Delta have resulted in pollution of much of the region's vegetation, fishponds and drinking water directly impacting indigenous communities. Steiner (2010, p. 4) identifies that over the course of 50 years of oil production in the Niger Delta the region has suffered 'extensive habitat degradation, forest clearing, toxic discharges, dredging and filling, and significant alteration by extensive road and pipeline construction from the petroleum industry'. Yet, the major oil companies operating in Nigeria generally embrace the concepts of ethical operations and human rights compliance often doing so by claiming to implement these in their corporate social responsibility (CSR) policies. Yet, the extent to which they do so, the content of these policies and their applicability to the concepts of environmental compliance varies considerably. While there has been widespread adoption of CSR policies by businesses in developed countries in the last 20 years, business activities are often not subject to international law or human rights norms. As a result, CSR is largely voluntary and while CSR and CER are routinely embraced by business, the question remains as to what extent this is just business proclaiming what people expect it to do while at the same time continuing to act either illegally or through CSR and Public Relations (PR) discourse downplaying the harm caused by business practices. Many corporations in the oil and gas industry such as Shell and BP have good CER policies on paper, but their activities result in environmental harms that a green criminological perspective would argue should be incorporated within a widened environmental crime definition (White 2008; White and Heckenberg 2014). Situ and Emmons (2000) identify that environmental 'crime' is predominantly a civil matter; in other words, fines and administrative penalties are the main technique for dealing with environmental wrongdoing. The lack of effective international law and the reliance on national (state) legislators to define what environmental crime is result in it often not being seen as a priority criminal justice issue. Thus, the issue for green criminology to consider is how should good standards of CER be enforced and whether solely defining what is legal or illegal suffices or should CER go further; including regulated standards of behaviour and expectations of corporate behaviour that the public will accept as ethical compliance?

A number of international environmental conventions require states to provide for effective environmental protection. Brown Weiss (1993) identified that until the 1960s environmental issues were viewed as state concerns and there was a lack of appreciation of the need for international environmental agreements. The *Convention on International Trade in Endangered Species of Wild Fauna and Flora* (CITES) was one of the first and oldest international legal agreements on environmental issues; it provided a framework for future wildlife and animal protection measures (Zimmerman 2003). It is, however, worth noting that in 1982 the UN General Assembly adopted a World Charter for Nature which contains the following five principles of conservation:

1. Nature shall be respected and its essential processes should be unimpaired.
2. Population levels of wild and domesticated species should be at least sufficient for their survival and habitats should be safeguarded to ensure this.
3. Special protection should be given to the habitats of rare and endangered species and the five principles of conservation should apply to all areas of land and sea.
4. Man's utilization of land and marine resources should be sustainable and should not endanger the integrity or survival of other species.
5. Nature shall be secured against degradation caused by warfare or other hostile activities.

In principle, the UN Charter provides a mechanism for protecting the environment (and animals) from harm by providing a conservation framework that requires active protection of nature. However, in practice, implementation of the charter (and subsequent international law initiatives) relies on national environmental legislation. This is problematic in relation to corporate activity as the response to non-compliance committed by legitimate corporate actors rarely engages the criminal law.

The reality is that countries have different laws and 'frequently quite different approaches to dealing with environmental crime' (White 2007, p. 184) compounded by much environmental damage being committed by corporations who fall outside the remit of much criminal law. In many countries, environmental crime falls within the jurisdiction of the enforcement arm of the state environment department, rather than being integrated into mainstream criminal justice. Thus, Western conceptions of environmentalism, the need to protect the planet and criminalise environmentally damaging actions are not universally shared and the concept of corporate criminal environmental activity is a relatively new concept. Corporate environmental responsibility and human rights will often overlap although this is an area where MNCs are not routinely bound by international norms. Voiculescu and Yanacopulos (2011, p. 4) identify the UN as being at the forefront of devising universally acceptable standards to embed 'respect for human rights norms and abstention from corrupt practices' into business and transnational corporations' operating practices. Harvard Professor, John Ruggie, the special representative to the UN Human Rights Council, identified that 'the failure to enforce existing laws that directly or indirectly regulate business respect for human rights is often a significant legal gap in State practice' (United Nations Human Rights Council 2011, p. 8). Thus, not only is judging appropriate standards of corporate behaviour problematic but so is enforcing such standards and remedying environmental/human rights problems caused by corporations. Business activity that harms or impacts on the environmental rights of communities is subject to a mixture of voluntary compliance, regulatory activity and victim litigation primarily driven by national legislation. Quinney's (1970) idea of crime as a social construction identified that acts defined as crime are, for the most part, behaviours undertaken by relatively powerless social actors. But green criminology has also focussed on harms such as corporate environmental harms as 'crimes of the powerful' (Walters et al. 2013). The reality being that while there is no single accepted and binding CER standard, there is a need to consider CER as not just dealing with *actual* corporate illegality but also with environmental harms caused by MNC activities even where legal, especially in the numerous cases where corporations fail to remedy the environmental harm they have caused (Alcock and Conde 2005).

Harris (2011, p. 39) highlights that corporations may adopt CSR/CER for a variety of reasons, principally:

1. Acting ethically is the right way for the company to behave
2. Doing what is right and fair is expected of an organisation
3. Acting ethically is in the organisation's best interests.

The extent to which one (or all) of these reasons applies and provides the corporation's motivation for integrating CER can have an effect on whether CER is adopted as part of operational practices ensuring that the corporation considers the impact and wider implications of its activities, or whether CER becomes solely an aspect of marketing and brand management or simply a PR tool. Amnesty International (2009, p. 57) argues that international standards exist for oil industries that MNCs operating in the Niger Delta should be well aware of. Most oil companies operating in Nigeria have in place policies which commit them to good practice in terms of environmental and social impacts. However, Amnesty's research concluded that serious failures existed in some companies' actions when it came to identifying and preventing pollution and environmental harm that leads to human rights abuses. The reasons given included failings because of: 'a significant gap between policy and practice; lack of effective oversight; or through under-funding' (Amnesty International 2009, p. 57). Controversially, Amnesty (2009, p. 57) concluded that the practice of some companies 'went beyond failure to ensure that their operations do not harm the environment, and could not be viewed as anything other than acts of negligence'. Thus, Amnesty identified that the principle of CER and adoption of CER policies is not reflected in field practice.

Multinational Oil Corporations and Environmental Harms in Africa

Oil extraction in Africa and MNCs behaviour typifies a number of elements common to corporate crime discourse and the green criminological perspective on environmental harm. Amao (2011, p. 1993) identifies that Nigeria's oil and gas revenues account for 95 % of the country's export earnings, over 40 % of its GDP and around 80 % of Government revenue. Oil is thus integral to Nigeria's economy and the lives of its citizens and the actions of MNCs involved in the oil industry have the potential to impact not just the environment but also the human rights of Nigerian citizens. Amnesty International and CEHRD (2011, p. 33) identify that Nigeria's oil regulation system contains laws and regulations that prohibit the pollution of land and water and which also require oil companies to promote good oil field practice and comply with international standards. However, enforcement of these regulations is considered to be ineffective. Baumüller et al. (2011, p. 1) noted that 'negative impacts of the oil industry are a major concern in sub-Saharan Africa (SSA), threatening not only the health of local communities but also the livelihoods they depend on'. They outline that 'consequences attributed to oil spills and gas flaring include the collapse of local fishing and farming, the loss of habitat and biodiversity, acid rain damage and health impacts of air, noise and light pollution' (Baumüller et al. 2011, p. 17). Such impacts are described as 'chronic and cumulative' and result in 'a severely impaired coastal ecosystem and compromised livelihoods and health of the region's impoverished residents' (FMoE et al. 2006, p. 2). The reality is that decades of oil exploration in the Niger Delta have resulted in pollution of much of the region's vegetation, fishponds and drinking water and in some cases have lowered the life expectancy of citizens in some part of the region. The UN Environment programme's (UNEP 2011, p. 167) assessment concluded that 'oil spills are frequent events in Ogoniland'. Oil spills negatively impact the Niger Delta communities destroying livelihoods and endangering the health of local populations to the extent that 'oil contamination in Ogoniland is widespread and severely impacting many components of the environment' (UNEP 2011, p. 204). Concluding that the effects of oil pollution are felt by the Ogoni people 365 days of the year, UNEP (2011, p. 204) also commented that given the low average life expectancy of under 50 years in Nigeria 'it is a fair assumption that most members of the current Ogoniland community have lived with chronic oil pollution throughout their lives'.

White (2012, p. 15) identifies that 'much environmental harm is intrinsically transnational' and is by its very nature mobile and easily subject to transference. The global reach of MNCs is situated within international markets and systems of production, requiring a system of understanding and addressing environmental harm that incorporates appreciation of its international dimensions (Beirne and South 2007). In this respect, where oil companies cause environmental harms in developing world countries, their activities may be viewed as a crime of the powerful (Brockington et al. 2008; Tombs and Whyte 2003), irrespective of whether formally recognised as such by the criminal law or whether civil law action is required to address the harm. Nieuwoudt (2007) has suggested that bribery by MNCs in seeking to win exclusive drilling and processing or prospecting licences has resulted in payments to the overseas accounts of government ministers or payment of such things as their children's private school fees. Thus, the oil extraction industry raises a number of white collar and corporate crime concerns.

Oil Extraction and Criminological Concerns

From a green criminological perspective, oil extraction in Nigeria illuminates a number of concerns not least those of environmental justice and associated issues of social inequality and the marginalisation of indigenous communities (White 2007). While much activity of criminological concern is likely regulatory non-compliance rather than actual crime, allegations linking the activities of western oil companies to perceived illegality have been made and are further discussed later in this chapter. In the context of green criminology's focus on environmental harms rather than crimes, the demand and wealth of western oil consumers provide a strong driver for maximising profits by minimising regulatory compliance and the integration of local people into the profits of natural resources. Contrary to Baumol's (1990) notion of a distinct split between legal and illegal entities, legal entities may engage in non-compliance as a legitimate profit-enhancing business tactic and legal entities may operate alongside illegal ones. In this context, lawbreaking related to misuse of natural resources represents an environmental injustice (Pearce and Tombs 1998), and a crime of the powerful (Tombs and Whyte 2003). Arguably, where oil extraction has a negative effect on local ecosystems and wildlife, this also represents a species justice concern.

From a criminological perspective, the corporate response to problems within the oil extraction industries indicates Sykes and Matza's (1957) neutralization techniques in operation. Given the continued legality of oil extraction and natural resource exploitation and sustainable use, MNCs engage in lawful activity that is then linked to perceived wrongdoing in a way that arguably challenges their status as legitimate actors by labelling certain actions as 'criminal', requires a response. One consequence of this is that the MNCs respond to the labelling either in an increasingly deviant manner (McBarnet 2006; Sykes and Matza 1957) or via the use of neutralisation techniques which assert their core values and the underlying legitimacy of their actions while challenging the enforcement or regulatory regime. Thus techniques such as denial of responsibility, claim of entitlement, denial of the necessity of the law, defence of necessity and condemnation of the condemners are all present reflecting the research of Sykes and Matza (1957) which identified that individuals involved in crime use these techniques both before and after engaging in illegal activity. Applied to the actions of MNCs in Nigeria where NGOs and civil society play a significant role in pursuing legal action, denial of the right of NGOs to act as 'enforcers' and of the matter as being worthy of court action rather than regulatory control are legitimate tactics before and during any proceedings. A complete list of possible neutralizations employed by corporate offenders (as with some other offenders) can be outlined as follows:

1. The denial of responsibility
2. The denial of injury or the existence of a recognised 'victim'
3. The condemnation of the condemners
4. The appeal to higher loyalties
5. The defence of necessity
6. The denial of the necessity of the law
7. The claim of entitlement

The precise techniques and neutralisations used are dependent upon the specific situation involved and may also be subject to different motivations on the part of MNCs. However, particularly when legal action is initiated or public condemnation of MNC actions is at play, evidence of the use of neutralisation techniques is often easy to ascertain. Using some case examples relating to legal action against MNCs, the following section discusses these issues in more detail.

Legal Action Against Oil Companies for Environmental Harms and Human Rights Abuses

As mentioned above, despite the existence of various conventions and national legislation which often provides for environmental protection on paper, the regulatory approach and criminal justice response to corporate environmental crime is often limited. Stallworthy (2008) identifies a conflict between private interests (such as those of profit-making companies) and the perceived public good of environmental protection. Lynch and Stretesky (2014) also identify that the power of neoliberal markets is such that an unwillingness to view legitimate market actors (e.g. MNCs) as the subjects of the criminal law is a widespread notion that impacts negatively on the ability of the criminal law to deal with corporate environmental non-compliance (Dybing 2012). Corporate environmental offending or non-compliance is often not a core policing responsibility and is frequently left to NGOs to monitor and pursue enforcement action (Nurse 2013) or is dealt with via 'lesser' options like environmental tribunals and is the responsibility of environment departments like DEFRA—the Department of Environment, Food and Rural Affairs (in the UK)—rather than criminal justice ones like the Home Office.

UNEP identifies that there are considerable difficulties with bringing legal action in Nigeria using domestic legislation due to the inaccessibility of much legislation.

Articulating the problems inherent in effective use of the legal system, UNEP (2011, p. 142) states that 'few texts are available online and many are not easily available even in paper form. In addition, printed copies of legislation, such as the 'Laws of the Federation of Nigeria', are extremely expensive and therefore limited to those able to bear the costs'. Thus, there is a lack of transparency within the legal system which undermines both confidence in its efficacy and in the ability of marginalised citizens to access the system and gain appropriate remedies. Even where legal action can be taken, allegations of judicial corruption and of government interests in the continued operation of MNCs as significant contributors to the economy cast doubt on the success of legal proceedings although various efforts have been attempted (Frynas 2004).

Despite these concerns, a number of legal cases have been brought in recent years which indicate that some form of ecological justice (Benton 1998; Lynch and Stretesky 2014) allied to a concern with effective human rights justice is possible. Amnesty International (2009) identify that in June 2009 a settlement was reached between Shell and the relatives of writer and human rights campaigner Ken Saro-Wiwa and eight other members of the Ogoni people who were executed by the Nigerian State in 1995. Various commentators have linked the executions to Saro-Wiwa's environmental and human rights advocacy which arguably represented a direct challenge to continued oil exploration and 'brought Nigeria and the international oil companies that operate there into international disrepute' (Baumuller et al. 2011, p. 17). Amnesty (2009, p. 4) confirms that Shell agreed to pay the plaintiffs US$ 15.5 million (£ 9.7 million) as a humanitarian gesture while not admitting liability and reiterating that Shell had no part in the violence in Ogoniland, citing the payment as a compassionate payment to the plaintiffs and the estates they represent in recognition of tragic events in Ogoni land. The denial of responsibility in this case while given support by the lack of any criminal trial of oil industry staff also represents an appeal to higher loyalties and the continued operation of the oil industry.

In March 2012, a group of approximately 11,000 Nigerians from Bodo, Ogoniland, launched a law suit against Royal Dutch Shell at the London High Court, seeking tens of millions of dollars in compensation for two oil spills in 2008. The case (still unresolved at the time of writing) illustrates denial of responsibility as an initial neutralisation technique, blaming the harm on sabotage, thus dismissing the MNCs alleged liability. Vidal (2014a) however, notes that Shell's Nigerian subsidiary, Shell Petroleum Development Company of Nigeria (SPDC), accepted responsibility and after several years of negotiation offered £ 30 million to compensate nearly 15,000 people who had claimed damages. However, the compensation amount was rejected by community leaders as inadequate with Vidal (2014a) reporting that the communities allegedly claimed damages of around £ 300 million. Subsequent evidence brought before the courts suggests that denial of injury and denial of responsibility combined when Shell's investigation reports claimed that only about 4000 barrels of oil were spilt in total whereas an independent assessment published by a US firm Accufacts Inc. calculated that the amount spilt exceeded 100,000 barrels (Vidal 2014b). While Shell acknowledged errors in its assessment

(Vidal 2014b), denying the scale of the spill potentially has the effect of minimising the perceived damage and injury caused to the community.

In a case finally subject to judgment in January 2013, *Milieudefensie* (Friends of the Earth the Netherlands) and four Nigerian farmers brought charges against both Shell Nigeria (SPDC, a subsidiary of Shell) and the Dutch parent company/multinational (Shell PC) due to oil pollution. The case, unique in being the first time a Dutch multinational has been brought before the court in its own country for environmental damage caused abroad, focuses on three oil leaks in the villages of Goi, Aruma and Ikot Ada Udo. *Milieudefensie* and the four victims argued that Shell should repair the damage in the three villages by properly cleaning up the oil, prevent new leaks from occurring in the future, by properly maintaining its material and protecting it from sabotage, and compensate the four victims for the damages suffered. The case sought to determine Shell Headquarters in The Hague as responsible for damages caused by its subsidiary SPDC in Nigeria, establishing liability on the part of corporations for the behaviour of their subsidiaries. Illustrating the neutralisation techniques of condemning the condemners and denial of the necessity of the law, Shell challenged the jurisdiction of the Dutch Court to hear the claims. However, despite Shell's objections, the District Court ruled that it was justified to adjudicate on the lawsuits against all Shell entities in The Netherlands, because the lawsuits were closely connected, applying an EU law principle where EU and overseas cases can be heard together in the interests of justice. The Hague District Court found that four out of five oil spills, which were the subject of the action, were not the result of poor maintenance by Shell, but were caused by sabotage by third parties. The court applied the relevant Nigerian law which specifies that an oil company is in principle not liable for any oil spills resulting from sabotage; the basis being that this is not an action of the oil company, but results from damage or environmental harm caused by a third party. Accordingly, the court dismissed four out of the five cases and, in respect of four lawsuits regarding an oil spill near the village of Goi in 2004 and an oil spill near the village of Oruma in 2005, the District Court concluded that Shell Nigeria took sufficient measures to prevent sabotage of its submerged oil pipelines. However, in the fifth claim relating to two oil spills in 2006 and 2007 from an abandoned wellhead near the village of Ikot Ada Udo, the District Court concluded that Shell Nigeria has had violated its 'duty of care' under applicable Nigerian law and had committed the 'tort of negligence'. The District Court acknowledged that the 2006 and 2007 spills were also the result of an act of sabotage but noted that this was committed in a very simple way near that village by using a wrench to remove above-ground heads of an oil well abandoned by Shell Nigeria. The court ruled that Shell Nigeria could and should have easily prevented the sabotage by installing a concrete plug prior to 2006, whereas it only did so in 2010 while the lawsuit was pending. The court therefore concluded that Shell could be held *partly* responsible for pollution in the Niger Delta. The judgement, while falling some way short of declaring Shell responsible for all of the pollution incidents, identified that Shell should have prevented sabotage at one of its facilities in Nigeria which caused a spill damaging to the environment and local wildlife resources.

Remedying the Harm and Enforcing CER

Corporations who break environmental laws and accepted standards of behaviour are sometimes subject only to civil or administrative sanctions. Baumüller et al. (2011, p. 7) identified that while oil companies will sometimes take action to address the harmful impacts of their activities, CSR activities 'largely remain piecemeal and short term, community engagement is inadequate and requirements for accountability and transparency are either insufficient or not enforced'. Amnesty International (2011) points out that in the Niger Delta, oil companies frequently take the lead in the investigations process rather than regulators, and suggests that the dominant role of oil companies in the investigation process creates a conflict of interest. While oil companies are obliged to clean up oil spills, any spills that are the result of sabotage do not attract compensation payments because compensation is only paid if the cause of a spill is judged to be the company's fault (for example due to equipment failure). Thus, as the cases discussed earlier in this chapter highlight, companies are able to deny responsibility where sabotage is the cause of an incident with the result that claims of sabotage remain hotly contested (Amnesty International 2009 and 2011). Ayanruoh (2009) argues that a fundamental and systemic disregard for human rights and the inequities of resource allocation exists in Nigeria. Thus 'a legal regime requiring the regulation of corporations, rather than states or individuals, is necessary to address the nature of corporations as actors in the human rights field' (Ayanruoh 2009). However, in some respects, international law has already effectively recognized duties of corporations. Slapper (2011, p. 95) identifies that there have been modest developments in the use of the civil law to address corporate abuses and that 'apart from a growth in domestic criminal liability of corporations' there has been an increase not only in civil litigation against companies 'but also the advent of domestic liability for corporate torts that are committed abroad'. Thus, while international law may not yet have caught up with corporate abuses, domestic law might, in some cases, provide a civil remedy.

Slapper's (2011) point is illustrated by US civil law in the form of the *Alien Tort Claims Act 1789*, which in principle allows action to be taken against companies for their deeds overseas. The *act* confers on US federal courts jurisdiction over 'any civil action by an alien for a tort only, committed in violation of the law of nations or a treaty of the United States'. Thus, where corporate acts, which are the subject of litigation, raise international concerns and constitute a crime against humanity, a remedy potentially exists for victims of corporate abuses able to bring a case in US courts. Nigerian plaintiffs brought an action in the US courts under the Alien Tort Statute (ATS) in *Kiobel v. Royal Dutch Shell Petroleum Co., et al.* The plaintiffs in *Kiobel*, all Nigerian citizens from the Ogoni region, brought claims under the ATS alleging the corporate defendants (effectively Shell and its associated companies) aided and abetted human rights abuses committed by the Nigerian government, including torture; cruel, inhuman and degrading treatment; and arbitrary arrest and detention. The case was reviewed by the US Supreme Court that examined both the availability of corporate liability under the ATS and the statute's extraterritorial reach. On 17 April 2013, the Supreme Court issued its decision in the case by way of a majority decision which found that there is a presumption against extraterritorial application of the ATS, and that presumption can only be overcome when the matter 'touches and concerns' the USA with 'sufficient force'. Stohr (2013) argued that the Supreme Court's decision effectively insulated MNCs from some lawsuits over activities abroad because all of the actions complained of took place in Nigeria. The implication, explained further in the three concurring opinions, is that unless the underlying actions also take place in the USA (in *Kiobel* these were predominantly human rights abuses) it will be difficult to succeed in the US Courts. Denny (2013) explains that 'jurisdiction over foreign subsidiaries of US corporations is precluded unless the subsidiary acts as the alter ego of the parent under its direct control, or independently maintains sufficient jurisdictional contacts with the forum'. This does not, however, mean that no remedy exists under the ATS or that cases cannot be brought, just that a clear link between corporate actions in the USA must also be established. The finding in *Kiobel* is in some ways consistent with EU law where cases can be brought against a parent company resident in the EU where it can be shown that the relevant management decisions (i.e. those which influenced or caused the local incident) were made at parent company level and thus effectively the harmful decisions (or at least some part of them) were made within the EU. However, as outlined earlier, the US test is somewhat stricter.

Fagan and Thompson identify class actions as being the primary legal mechanism feared by US corporations, which Hodges (2008, p. 2) identifies as being based on a model where 'one individual claim is asserted to represent a class of others, whose owners are bound by the result of the single claim unless they opt-out of the class and procedure'. The class action procedure allows for punitive damages and requires parties to meet their own costs (Fagan and Thompson 2009, p. 56–57). Connelly and Smith (1999) suggest that collective action is often the necessary solution to environmental problems where civil action can be used to seek a remedy in ways that criminal action often fails to. The failures in formal (state) enforcement action in Nigeria have required NGOs, such as Friends of the Earth, to become involved in cases, often using civil class action mechanisms to bring action for aggrieved citizens.

Some Conclusions on Oil Companies and Environmental Justice

The cases and examples of non-compliance discussed within this chapter illustrate the difficulty of achieving effective environmental justice where MNCs are able to exert considerable economic power in developing countries. Such cases also demonstrate the difficulties facing communities attempting to enforce environmental laws where the domestic judicial system, legislature and regulatory enforcement mechanisms are perceived as inadequate to deal with human rights and environmental abuses committed by MNCs and where, strictly speaking, corporate environmental non-compliance may not be classified as criminal. However, the Shell case in the Netherlands is unique in being the first time a European ‘home country’ court has exercised jurisdiction to consider environmental damage caused abroad. It is also the first time Shell has been ‘ordered’ to pay compensation for damage caused in Nigeria.

The impact of the Shell case is difficult to quantify. Initial responses to the case saw both sides claiming victory with Shell declaring itself satisfied with the verdict and commenting that it would not change the way it conducts its business in Nigeria (UPI 2013). Shell also successfully defended claims against its parent company for the actions of its Nigerian subsidiary. However, Friends of the Earth and the Nigerian Plaintiffs also claimed victory because despite the dismissal of four claims, Shell was found liable in a Dutch Court for failings in Africa. Given that the plaintiff, the Nigerian defendant and the environmental claims at the heart of the case all took place outside of the EU, the ruling can be seen as a significant step forward in allowing court action to be taken against companies for their actions overseas. Denny (2013) argues that the judgments of the District Court of The Hague can thus be seen as part of the development of international law dealing with corporate harms, which allows multinationals to be held accountable in their home jurisdiction for ‘damage that has been caused by or is related to the operations of their foreign subsidiaries’. This is significant in part because the use of subsidiaries in countries with (perceived) weak judicial systems is a mechanism through which corporations seek to protect themselves from costly environmental litigation (O’Hara 1990; Namballa 2014). As discussed earlier in this chapter, victims of environmental harm may seek to bring proceedings before an EU or US court rather than in their local jurisdiction for various reasons. From a green criminological perspective, the dictates of effective ecological justice are such that the most effective forum in which to provide environmental justice and raise the profile of environmental justice is the one that should be used. Thus, a valid additional imperative for organisations such as Friends of the Earth Netherlands (Milieudefensie) to bring a claim against both the subsidiary and the parent company in the parent company’s western home country is to draw attention to the issue of CER within western markets.

Although in the Shell Nigeria case, direct liability of the parent was not established, multinationals that have their seat or headquarters in a western jurisdiction continue to run a serious risk of being sued for damages caused or related to operations of their foreign subsidiaries. The publicity generated by such claims is high, and this seems exactly what non-governmental organisations such as Friends of the Earth and Greenpeace are likely to aim for, consistent with their roles in using the law for both campaigning and enforcement purposes (Nurse 2013). From an ecological justice perspective, while transnational torts against parent companies and their foreign subsidiaries are complex, costly and lengthy, they offer the hope of incorporating green criminological perspectives of justice for non-human nature into justice systems. Notwithstanding the difficulties of relying on NGOs to facilitate legal action, that these may be civil rather than criminal is irrelevant if some form of justice can be achieved.

References

Alcock, R., & Conde, C. (2005). Socially and environmentally responsible business practices: An Australian perspective. *Corporate Governance Law Review, 1,* 329–388.

Amao, O. (2011). Human rights, ethics and international business: the case of Nigeria? In A. Voiculescu & H. Yanacopulos (Eds.), *The business of human rights: An evolving agenda for corporate responsibility*. London: Zed Books/The Open University.

Amnesty International. (2009). *Nigeria: Petroleum, pollution, and poverty in the Niger Delta.* London: Amnesty International Publications

Amnesty International and CEHRD. (2011). *The true 'Tragedy' delays and failures in tackling oil spills in the Niger Delta.* London: Amnesty International Publications

Ayanruoh, F. E. (2009). 'Corporate responsibility and human rights abuse in the Niger Delta,' Global Policy Forum November 2009. https://www.globalpolicy.org/security-council/dark-side-of-natural-resources/oil-and-natural-gas-in-conflict/africa/48470-corporate-responsibility-and-human-rights-abuse-in-the-niger-delta.html. Accessed 15 Dec 2014.

Baumol, J. (1990). Entrepreneurship: Productive, unproductive and destructive. *Journal of Political Economy, 98*(5), 893–921.

Baumüller, H., Donnelly, E., Vines, A., & Weimer, M. (2011). *The effects of oil companies' activities on the environment, health and development in Sub-Saharan Africa*. London: Chatham House.

Beirne, P., & South, N. (Eds.). (2007). *Issues in green criminology: Confronting harms against environments, humanity and other animals*. Devon: Willan Publishing.

Benton, T. (1998). Rights and justice on a shared planet: More rights or new relations?. *Theoretical Criminology, 2*(2), 149–175.

Brockington, D., Duffy. R., & Igoe, J. (2008). *Nature unbound: Capitalism and the future of protected areas*. Abingdon: Routledge.

Brown Weiss, E. (1993). International environmental law: Contemporary issues and the emergence of a new world order. Georgetown *Law Journal, 81,* 675–697.

Connelly, J., & Smith, G. (1999). *Politics and the environment: From theory to practice*. London: Routledge.

Denny, A. (2013). *The Shell Nigeria cases—an important precedent for transnational liability claims*. London: Allen and Overy. http://www.allenovery.com/publications/en-gb/Pages/The-Shell-Nigeria-cases-an-important-precedent-for-transnational-liability-claims.aspx. Accessed 21 Dec 2014.

Doyon, J. (2014). Corporate environmental crime in the electronic waste industry: The case of executive recycling, inc. In A. Nurse (Ed.), *Critical perspectives on green criminology* (Internet Journal of Criminology). Nottingham: New University Press/Flashmouse Publishing Ltd.

Dybing, S. S. (2012). Environmental harm: Shifting causes and shifting legislative dynamics. In R. Ellefsen, R. Sollund, & G. Larsen (Eds.), *Eco-global crimes: Contemporary problems and future challenges*. Farnham: Ashgate.

Fagan, N., & Thompson, L. (2009). Corporate responsibility and group redress mechanisms. *Business Law International, 10*(1), 51–60.

FMoE, NCF, WWF, & CEESP-IUCN. (2006). *Niger Delta natural resource damage assessment and restoration project Phase 1—Scoping report.* Federal Ministry of Environment, Nigeria Conservation Foundation, WWF UK and CEESP-IUCN Commission on Environmental, Economic and Social Policy.

Frynas, J. G. (2004). Social and environmental litigation against transnational firms in Africa. *The Journal of Modern African Studies, 42,* 363–388. doi:10.1017/S0022278X04000230.

Harris, F. (2011). Brands corporate social responsibility and reputation management. In A. Voiculescu & H. Yanacopulos (Eds.), *The business of human rights: An evolving agenda for corporate responsibility*. London: Zed Books/The Open University.

Hodges, C. (2008). *The reform of class and representative actions in European legal systems: A new framework for collective redress in Europe*. Oxford: Hart Publishing.

Lynch, M. J., & Stretesky, P. (2014). *Exploring green criminology: Towards a green criminological revolution*. Farnham: Ashgate.
McBarnet, D. (2006). After Enron will 'whiter than white collar crime' still wash? *British Journal of Criminology, 46,* 1091–1109.
McGrew, A. (Ed.). (1997). *The transformation of democracy?: Democratic politics in the new world order*. Milton Keynes: The Open University.
Namballa, V. C. (2014). Global environmental liability: Multinational corporations under scrutiny. *Exchanges: The Warwick Research Journal,* 1(2), 1–25. (ISSN 2053–9665, S. 1.). http://exchanges.warwick.ac.uk/index.php/exchanges/article/view/26/100. Accessed 20 Dec 2014.
Nieuwoudt, S. (2007). How to Turn the Curse of Oil into a Blessing, Inter Press Service News Agency. http://www.ipsnews.net/2007/05/trade-africa-how-to-turn-the-curse-of-oil-into-a-blessing/. Accessed 31 Aug 2014.
Nurse, A. (2013). *Animal harm: Perspectives on why people harm and kill animals*. Farnham: Ashgate.
Nurse, A. (2014). Cleaning Up greenwash: The case for enforcing corporate environmental responsibility. In A. Nurse (Ed.), *Critical perspectives on green criminology* (Internet Journal of Criminology). Nottingham: New University Press/Flashmouse Publishing Ltd.
O'Hara, G. P. (1990). Minimizing exposure to environmental liabilities for corporate officers, directors, shareholders and successors. *Santa Clara High Technology Law Journal, 6*(1), 1–22.
Pearce, F., & Tombs, S. (1998). *Toxic capitalism: Corporate crime and the chemical industry*. Aldershot: Dartmouth.
Quinney, R. (1970). *The social reality of crime*. Boston: Little Brown.
Situ, Y., & Emmons, D. (2000). *Environmental crime: The criminal justice system's role in protecting the environment*. Thousand Oaks: Sage.
Slapper, G. (2011). Violent corporate crime, corporate social responsibility and human rights. In A. Voiculescu & H. Yanacopulos (Eds.), *The business of human rights: An evolving agenda for corporate responsibility*. London: Zed Books/The Open University.
Sollund, R. (2012). Introduction. In R. Ellefsen, R. Sollund, & G. Larsen (Eds.), *Eco-global crimes: Contemporary problems and future challenges*. Farnham: Ashgate.
Stallworthy, M. (2008). *Understanding environmental law*. London: Sweet and Maxwell.
Steiner, R. (2010). *Double standard: Shell practices in Nigeria compared with international standards to prevent and control, pipeline oil spills and the deepwater horizon oil spill*. Amsterdam: Friends of the Earth Netherlands.
Stohr, G. (2013). Companies shielded as US Court Cuts Human-Rights Suits, *Bloomberd April 2011*. http://www.bloomberg.com/news/2013-04-17/companies-get-shield-as-top-u-s-court-curbs-human-rights-suits.html. Accessed 15 Dec 2014.
Sykes, G. M., & Matza, D. (1957). Techniques of neutralization: A theory of delinquency. *American Sociological Review, 22,* 664–673.
Tombs, S., & Whyte, D. (2003). Unmasking the crimes of the powerful. *Critical Criminology, 11*(3), 217–236.
United Nations Environment Programme. (2011). *Environmental assessment of Ongoniland*. Kenya: United Nations Environment Programme. http://postconflict.unep.ch/publications/OEA/UNEP_OEA.pdf. Accessed 14 Dec 2014.
United Nations Human Rights Council. (2011). *Guiding principles on business and human rights: Implementing the United Nations "Protect, Respect and Remedy" framework*. New York: United Nations. http://www.business-humanrights.org/media/documents/ruggie/ruggie-guiding-principles-21-mar-2011.pdf. Accessed 20 Dec 2014.
UPI. (2013). Shell OK with Nigerian spill Verdict. http://www.upi.com/Business_News/Energy-Resources/2013/01/30/Shell-OK-with-Nigerian-oil-spill-verdict/UPI-13501359553335/. Accessed 10 June 2013.
Vidal, J. (2014a). Shell faces payouts in Nigerian spill case. http://www.theguardian.com/environment/2014/jun/20/shell-faces-payouts-nigerian-oil-spill-case. Accessed 20 Oct 2014.

Vidal, J. (2014b). Shell ignored internal warnings over Nigeria oil spills. http://www.theguardian.com/global-development/2014/nov/13/shell-nigeria-oil-spills-amnesty-international. Accessed 17 Dec 2014.

Walters, R., Westerhuis, D., & Wyatt, T. (Eds.). (2013). *Emerging issues in green criminology: Exploring power, justice and harm*. Basingstoke: Palgrave Macmillan.

White, R. (2007). Green criminology and the pursuit of social and ecological justice. In P. Beirne & N. South (Eds.), *Issues in green criminology: Confronting harms against environments, humanity and other animals*. Cullompton: Willan Publishing.

White, R. (2008). *Crimes against nature: Environmental criminology and ecological justice*. Cullompton: Willan Publishing.

White, R. (2012). The Foundations of Eco-global Criminology. In R. Ellefsen, R. Sollund, & G. Larsen (Eds.), *Eco-global crimes: Contemporary problems and future challenges*. Farnham: Ashgate.

White, R., & Heckenberg, D. (2014). *Green criminology: An introduction to the study of environmental harm*. London: Routledge.

Yanacopulos, H. (2011). Foundations—actors of change? In A. Voiculescu & H. Yanacopulos (Eds.), *The business of human rights: An evolving agenda for corporate responsibility*. London: Zed Books/The Open University.

Zimmerman, M. E. (2003). The black market for wildlife: Combating transnational organized crime in the illegal wildlife trade. *Vanderbilt Journal of Transnational Law, 36,* 1657–1689.

Dr. Angus Nurse is a senior lecturer in criminology at Middlesex University in London, UK.

Chapter 10
Legal and Extralegal Enforcement of Pollution by Seagoing Vessels

Judith van Erp, Toine Spapens, and Karin van Wingerde

Introduction

Seagoing vessels contribute considerably to marine and air pollution. The first reason is the dumping of waste containing oil at sea, which harms the marine environment as well as seabirds. The second is the heavy fuel oil (HFO) used as a propellant in seagoing ships. Whereas HFO is far from a clean product to begin with, hazardous chemicals and untreated waste oil blended with this type of oil is an important additional source of pollution.

This chapter first discusses the scope and nature of both activities. The first part of this chapter focuses on the dumping of waste oil at sea as well as on a case in which chemical waste created on board of the ship Probo Koala was dumped on land. Next, we address the blending of hazardous waste with HFO. Pollution caused by accidents and other activities at sea, for example from drilling, is excluded here.

Part 2 discusses the reasons why curbing these problems through law enforcement is often problematic. To begin with, although the aforementioned problems create serious environmental hazards they are still very difficult to detect, regulate, and monitor. In addition, polluters are often able to cleverly and creatively utilise regulatory loopholes in the grey zone between legal and harmful behaviour. This raises the question whether and how extralegal interventions, particularly social pressure, negative publicity, and 'naming and shaming', may contribute to the prevention of harms to the environment. Part 3 of this chapter addresses this question in more detail. The final part summarises and discusses the implications of the key findings.

J. van Erp (✉)
Utrecht University, Utrecht, The Netherlands
e-mail: j.g.vanerp@uu.nl

K. van Wingerde
Erasmus University Rotterdam, Rotterdam, The Netherlands

T. Spapens
Tilburg University, Tilburg, The Netherlands

T. Wyatt (ed.), *Hazardous Waste and Pollution*, DOI 10.1007/978-3-319-18081-6_10

Dumping of Waste Oil

The normal operating process of (seagoing) ships produces several types of waste containing oil. To begin with, ships running on HFO produce fuel oil sludge that cannot be used in the ship's engines. Sludge typically comprises 1.5–2 % of the fuel oil volume. Second, oil often leaks from engine and machinery spaces or from engine maintenance activities and mixes with water in the bilge, the lowest part of the hull of the ship. This so-called bilge water consists of a mixture of water, fuel, lubricating oil, solvents, and chemical waste. Finally, washing out the residue from the tanks of an oil tanker (crude oil washing) creates 'oily slops'. Ships are obliged to discharge the aforementioned types of waste in a port reception facility or treat it on board (e.g. burn it in an incinerator). However, dumping it at sea remains a cheaper method (Crist 2003).

Such dumpings have substantial effects on the environment. One reason is that they usually concentrate in shipping lanes, and the cumulative effect of dumpings, which itself may be relatively small, is substantial (National Research Council 2003, p. 28). Another is that a single dumping may accidentally hit a concentration of wildlife, such as birds, which happen to be in the wrong place at the wrong time (Camphuysen 2010; National Research Council 2003). Dumping of waste oil affects both birds and marine life and also pollutes the sediment threatening flora and fauna living on the sea floor (cf. Grontmij 2010, p. 38).

Ships for a longtime disposed of waste oil at sea. However, after the Second World War this increasingly polluted the coasts and already in 1954 countries convened in London and drew up the 'International Convention for the Prevention of Pollution of the Sea by Oil'. This treaty aimed primarily at better regulating the dumping of waste oil, but also obliged the governments to organise facilities for the reception of residues from oily ballast water and tank washings. Ships, however, were not yet obliged to dispose of these substances in ports. Reception facilities operated on a commercial basis and paid the ships for the waste oil that could be treated and reused.

A few years earlier in 1948, the United Nations had decided to establish an international organisation for marine safety. This body, the Inter-Governmental Maritime Consultative Organisation (IMCO), became operational in 1959 and pollution caused by ships was one of its priorities.[1] A conference organised in 1962 resulted in a number of amendments to the 1954 convention. Ships of over 20,000 gross register tonnage were no longer allowed to dispose of waste oil and oily substances at sea. These ships had to be fitted out with slop tanks in which the waste could be stored until it could be discharged at a port reception facility. In addition, dumping was no longer allowed in the entire North Sea and Baltic Sea. The amended convention entered into force in 1967, but soon afterwards several states urged to draw up an entirely new and more rigorous convention. In 1973, these efforts resulted in the 'International Convention for the Prevention of Pollution from Ships', better known

[1] In 1982, the name of IMCO was changed to International Maritime Organization (IMO).

as the MARPOL Convention. Its aim is to minimise marine pollution resulting from oil and all other hazardous substances as much as possible. Central to these efforts is a total ban on the dumping of waste at sea and the provision of harbour reception facilities where ships could dispose of it. The new convention entered into force in October 1983.

Despite the convention, ships still dump oily substances at sea (Vollaard 2013). One of the reasons is the fact that delivery at a harbour reception facility costs money. In a sector that suffers from cut-throat competition, any opportunity to save money presents an attractive proposition. One indicator for the dumping of oil is the percentage of common murre that die of oil pollution, because this bird is extremely vulnerable to oil. In the 1960s and 1970s, almost all of the dead common murre that washed up onto the Dutch North Sea coast were smeared with oil. Although the figure was greatly reduced in the years to follow, in the winter of 2011/2012, still about one third of the common murre found on the Dutch beaches had died of oil pollution (Camphuysen 2012). These findings show that the dumping of waste oil is an ongoing problem.

A Notorious Example: The Probo Koala Case

One notorious example where the illegal dumping of waste oil from a ship resulted in disaster is the Probo Koala case, although it is to a certain extent atypical. On the one hand, the waste was not the result of the ship's normal operations, but produced during an industrial process to upgrade low-quality oil which was carried out on board. On the other hand, the waste was not dumped at sea, but on land. In total, an estimated 450 tons of waste ended up in the countryside near Abidjan in the Ivory Coast (Amnesty International and Greenpeace Netherlands 2012).

The case had started in July 2006, when the Probo Koala arrived in the harbour of Amsterdam to dispose of the waste. The vessel was chartered by Trafigura, according to its website 'one of the world's leading independent commodity trading and logistics houses' .[2] Trafigura informed the harbour reception facility operated by Amsterdam Port Services (APS) that these were normal slops. When the Probo Koala arrived in the port, APS took samples of the waste and quickly discovered that it was not the 'mixture of gasoil, water, and caustic soda' that Trafigura had claimed it to be. APS notified the company that it could still accept the waste and take care of its treatment but that the costs would be € 900m, instead of € 20 that applied to normal slops (Amnesty International and Greenpeace Netherlands 2012; Spapens et al. 2013; van Wingerde 2015). Trafigura refused to accept this higher price, which amounted to an extra cost of several hundreds of thousands of Euros, and ordered APS to stop off-loading the waste and return it to the ship. Trafigura then found a local company in Ivory Coast, Compagnie Tommy, which agreed to the reception of the waste carried in the Probo Koala. Shortly afterwards the waste

[2] See: (Accessed 20 December 2014). www.trafigura.com.

dumped in the countryside, where the local community claimed that the waste caused injuries and even deaths to people who came into contact with it.

In March 2007, a similar problem manifested itself with the vessel Torm Thyra. Once again the waste was far more dangerous than normal slops and the fact that Trafigura had chartered the ship for this journey set off all alarms with the Dutch authorities.[3] In this case, all of the waste was discharged at the port of Amsterdam. Whether it was also treated correctly, however, was questioned in a Dutch documentary broadcast in October 2009.[4] Allegations were that the waste had been shipped to a Dutch company called North Refinery. This company was accused of being unable to treat the waste and that had mixed it with other chemical waste and sold it as a blend product for the production of HFO. This brings us to the second problem addressed in this chapter: 'dirty oil'.

'Dirty Oil'

Although various pollutant aspects of the oil industry related to the marine environment, such as oil drilling and oil spills, have received ample societal and scholarly attention (Jernelöv 2010),[5] the harmful effects related to HFO used by seagoing ship's engines have not been discussed widely. To begin with, seagoing vessels contribute substantially to air pollution because the burning of HFO releases sulphur dioxide and other pollutants into the air. In 2009, media reported that just 15 of the world's biggest ships may now emit as much pollution as all the 760 million cars in the world (Vidal 2009). Low-grade HFO has up to 2000 times the sulphur content of diesel fuel used in cars. Shipping is responsible for an estimated 18–30 % of the world's nitrogen oxide (NOx) pollution and 9 % of the global sulphur oxide (SOx) pollution. Furthermore, 70 % of all ship emissions are within 400 km of land (Vidal 2009). According to simulations, about 19 % of particulate matter emissions in Europe can be attributed to seagoing vessels (Andersson et al. 2009). The blending of untreated waste oil and chemical waste in HFO further contributes to the problem (Spapcns ct al. 2013).

The basis of HFO is residual oil, a rest product of refineries. This rest product as such is unfit for use in ship's engines unless it is blended with lighter fuels. Once, the manufacturers used diesel fuel for this purpose, but over the years, this became too expensive. An alternative were rest products of refineries, so-called cutter stocks. However, as refineries became more efficient there were less of these avail-

[3] Parliamentary proceedings, TK 2006–2007, 22 343, no. 178, p. 4.

[4] *Zembla*, 'De smerige olieroute', 11 October 2009, Available online at: http://tvblik.nl/zembla/de-smerige-olieroute (Accessed 5 February 2015).

[5] National Commission on the BP Deepwater Horizon Oil Spill and Offshore Drilling (2011). *Deep Water The Gulf Oil Disaster and the Future of Offshore Drilling. Report to the President.* http://www.gpo.gov/fdsys/pkg/GPO-OILCOMMISSION/pdf/GPO-OILCOMMISSION.pdf (Accessed 7 February 2015).

able. Fuel producers then turned to use light, flammable chemical waste as a blend product. There is basically nothing wrong with that as long as the chemical waste is properly treated before being reused. However, if this is omitted, this causes a number of problems.

To begin with, pollutants burn incompletely in ships' engines because the temperatures are not high enough and thus they end up in the atmosphere as particulate matter. Second, polluted HFO damages the ships engines and may even cause a shutdown. Ships not under control may cause serious problems when they run aground or collide with other ships or objects at sea. Third, the fuel can cause serious health problems for the crew. Stories can be heard of crewmembers that walk across the engine room vomiting or suffering burns from coming into contact with fuel.[6]

Inspections carried out from the 1980s onwards by Dutch authorities reveal that mixing untreated waste oil and chemical waste in HFO is common practice. From 2011 to 2013, the Dutch water police, the Human Environment and Transport Inspectorate, and the Netherlands Forensic Institute initiated operation 'Watchful'. During the first inspections, 21 of 28 ships required further investigation because of irregularities concerning bunker fuel ranging from environmental violations to tax evasion. Follow-up inspections rendered fewer results, not in the least because the ship's captains responded immediately when they received the word of police activity. Ships were observed to postpone entering the Dutch territorial waters until the inspections were concluded, for example. Criminal investigations revealed that because of the large quantities of fuel, profits easily amounted to millions of Euros in single cases.

In the port of Rotterdam, the trade in HFO is an important economic activity. Annually, over 20,000 ships refuel in Rotterdam, and the trade in HFO constitutes about one third of the trade in 'wet' products in the harbour (de Buck et al. 2011). In 2007, total deliveries amounted to about 13.4 million tons. In the same year, Furijah in the United Arab Emirates handled a similar amount and only Singapore (31.4 million tons) surpassed Rotterdam.[7] In the same year, the European ports of Antwerp (6 million tons), Gibraltar (4 million tons), and Piraeus (3 million tons) also appeared in the global top ten of bunker harbours.

There exist no international regulations for the blend components used to produce HFO. Only the end product must meet with certain qualifications. Annex VI of the MARPOL Convention sets limits to the sulphur oxide content of HFO and Regulation 18 specifically requires that fuel oil supplied to ships is to be free from inorganic acids or chemical wastes that could jeopardise the safety of the ship, be harmful to ships' personnel, or which would contribute overall to additional air pollution. These general requirements, however, left considerable room for legal discussion.

Theoretically, the Dutch legislation also allowed preventing the application of (chemical) waste as a blend component. However, these regulations only applied to 'waste' whereas suppliers consequently defined blend components as 'product', not

[6] *Zembla*, 'De smerige olieroute', 11 October 2009, Available online at: http://tvblik.nl/zembla/de-smerige-olieroute (Accessed 5 February 2015).

[7] See <www.bunkerworld.com>.

in the least because the producers of HFO did not have a license to process waste. National legislation about what is to be defined as waste differs widely between countries. This problem could only be solved by adopting a well-defined list of products that were allowed to be used as blend products.

Operation Watchful underlined that administrative bodies could not prevent fuel producers from using potentially harmful blend products. Furthermore, law enforcement agencies had great difficulty in finding the culprit when the end product violated existing regulations because the contamination could have occurred anywhere between the production of the fuel and the delivery to a ship's bunkers. The Dutch government responded to the questions of members of parliament that given the 'highly international nature of the market for HFO and the potential economic consequences for the Dutch traders' the International Maritime Organization (IMO) should decide upon additional regulations.

Legal Enforcement Problems

Traditionally, governments have sought to deal with the issues described earlier by increasing environmental regulation and enforcement. For instance, in the aftermath of the Probo Koala case, the European Union adopted a new directive, urging the member states to protect the environment through criminal law.[8] The main reason underlying this new directive was the belief that criminalisation and law enforcement are crucial to ensure compliance with environmental regulations. However, the strategy to influence the behaviour of firms using regulation and enforcement is filled with difficulty. Environmental regulation is often criticised for being inflexible, expensive for both firms and governments, and ineffective in shaping compliance behaviour of firms. This section discusses several ambiguities and legal complexities related to the pollution of seagoing vessels that create problems for detection and enforcement.

The Problem of Detection

As described earlier, one major factor contributing to the pollution by seagoing vessels is the economic incentive for dumping or blending waste containing oil at sea. Disposal of these wastes at specialised facilities is often time-consuming and expensive. Dumping waste at open sea or mixing it with HFO is simply cheaper. Apart from this cheaper method of disposal, blending waste products with HFO creates an extra incentive, because it also saves the costs of cleaner—and more

[8] Directive 2008/99/EC of 19 November 2008 on the protection of the environment through criminal law, official journal L328/28 of 6 December 2008.

expensive—fuel. In a recent study, Spapens et al. (2013, p. 187) estimated that this might result in an additional US $ 200–500 save per ton fuel oil.

While these economic incentives are significant, the dumping and blending of chemical waste products and residual oil are very difficult to detect for a number of reasons. One of the most widely used methods for monitoring pollution at sea is aerial surveillance (Lagring et al. 2012; Vollaard 2013). In the Netherlands, and in many other countries, the Coast Guard is responsible for monitoring and surveillance and makes surveillance flights approximately 7–8 times per week (Vollaard 2013, p. 16). The probability that oil spills are observed from the air is, however, small. Even though there are regular surveillance flights, Vollaard (2013, p. 18) shows that for most of the time (approximately 85 %) there is no aerial surveillance whatsoever. Moreover, the detection probability is even more limited because the aerial surveillance is difficult at night and during mist or fog. About 70 % of the surveillance flights, therefore, happen during the day, while most illegal activity occurs after sunset. As a result, the detection probability of an illegal dumping in the part of the North Sea monitored by the Dutch Coast Guard is estimated at roughly 1 % (Vollaard 2013, p. 19).

Another problem that limits the possibilities for detection revolves around the nature of the products involved. Waste oil and other hazardous waste materials are generally considered to be products that are highly vulnerable to manipulation and create opportunities for various forms of criminal manipulation (de Buck et al. 2011; Spapens et al. 2013; van Daele et al. 2007; White and Heckenberg 2011). For instance, most oil or waste spills at sea are undetectable within a few hours, because the products dissolve into the seawater rather quickly (Helmers 2002). Moreover, many oil products inhibit some of the same physical and chemical characteristics as many waste products, which makes blending of hazardous and nonhazardous wastes and blending of waste and oil less visible and harder to detect (Spapens et al. 2013, p. 115). Similarly, van Daele et al. (2007, pp. 35–36) emphasise the problem of so-called mirror entries. These define waste only as 'hazardous' when the concentration exceeds certain levels. By keeping the concentration under legal limits, the transportation, treatment and processing of these wastes remain largely under the regulatory radar.

However, even if the dumping and blending operations would be more visible and easier to detect, other complexities make it difficult to establish whether an actual violation has occurred, and subsequently to find the responsible actors and sufficient evidence to prosecute and convict them. A key issue in relation to the blending of waste and oil is how to distinguish between waste and reusable products (Gibbs et al. 2010; Spapens et al. 2013; van Daele et al. 2007; van Erp and Huisman 2010; White 2008). Blending residual oil with waste products creates a new lighter oil that can be used as a fuel for ships. The regulatory framework under the Basel Convention, however, allows for the possibility that waste materials that can be reused are not illegal by definition. In other words, the origin of the waste no longer determines its definition, but its use in practice. Nonetheless, this is often very difficult to assess because in different parts of the world, there are different values attached to products. It has been suggested to create a black list for products that are prohibited from blending into oil. However, the Dutch government has—rightfully—indicated that such a black list is merely symbolic in a sector in which purposeful law evasion is common practice, when it is not backed up by stringent enforcement.[9]

As with many other forms of environmental crime, the production processes of blending waste and oil are another source of complexity. There are often many steps and many different actors involved in this process, which makes it difficult to distinguish who is responsible and to establish a causal link between the producer of the chemical waste and the end product (Spapens et al. 2013, p. 115). Also, traditionally the blending of chemical products happened on land at a specialised refinery. Increasingly, however, firms find new locations to carry out these processes, including tanker terminals, but also at sea on board of a ship (Spapens et al. 2013, p. 181; van Wingerde 2015). This creates serious problems for oversight and makes it difficult to decide which country and authority has jurisdiction for enforcement. For instance, in the Probo Koala case, the ship was owned by a Greek firm; it was registered in Panama; it was chartered by a Dutch-based firm with its operating headquarters in London; and the ship sailed from Amsterdam to Estonia, Nigeria, and Ivory Coast (van Wingerde 2015). These difficulties also further impact the possibilities to prosecute, convict, and sentence offenders. Even when a violation is discovered, it is often very difficult to find sufficient evidence to bring the case before the court and to attribute the violations to the responsible actors.

The Problem of Enforcement: Too Big to Deter?

As the Probo Koala incident illustrates, the most prevalent response to many of the problems described here is to increase supervision and to impose harsher, more deterrent sanctions for violations. However, the deterrent effect of formal legal sanctions is often limited (Kagan et al. 2011; Thornton et al. 2005; van Wingerde 2012). Especially for large multinational business firms that are often involved in the trading and blending of oil products, penalties for environmental violations are insignificant compared to the profits of most of those firms. In the Probo Koala case, for example, Trafigura was sentenced in the Netherlands to pay a fine of 1 million € in 2011 while the company had its best year ever, with a turnover of US $ 1216 billion and a net profit close to US $ 1 billion.[10]

Power asymmetries and economic dependences between these large firms on the one hand, and governments and regulatory bodies, on the other, are often said to explain why dealing with corporate environmental crime is difficult (Hertz 2003; Snider 2010). Firms involved in the trading of commodities such as oil are among

[9] Dutch ministry of Safety and Justic, Letter to Parliament of 13 October 2013, reference number 434925, http://www.google.nl/url?sa=t&rct=j&q=&esrc=s&source=web&cd=9&ved=0CFkQFjAI&url=http%3A%2F%2Fwww.rijksoverheid.nl%2Fbestanden%2Fdocumenten-en-publicaties%2Fkamerstukken%2F2013%2F10%2F08%2Fstudie-vuile-olie%2Flp-v-j−0000004163.pdf&ei=jtDQVNqSCoKzUbWrg8AK&usg=AFQjCNHvWonNLieGVp4QWmkiFq-RIjdMeQ&bvm=bv.85076809, (Accessed 6 February 2015).

[10] Trafigura (n.d.). *Corporate brochure,* http://www.trafigura.com/useful_links/corporate_brochure.aspx (Accessed 24 May 2013), also http://www.ft.com/cms/s/0/99241c2c-2998-11e1=-8b1a-00144feabdc0.html#axzz2UClTHeoY (Accessed 24 May 2013).

the world's largest and most powerful firms. Oil is the biggest commodity market in the world (CEPS-ECMI Taskforce 2013, p. 74). The combined turnover of the ten largest commodity traders amounts to US $ 885 billion (Blas 2013). The four largest commodity traders, each have annual revenues of more than US $ 100 billion placing them in the same ranks as more familiar companies such as Apple (Shumsky and Kent 2014). The sales and profits of some of these firms are greater than the gross national product of many countries. For instance, the sales from the ten largest commodity traders are almost the same as the gross domestic product of Spain and South Korea. Moreover, the Dutch-based firm Vitol had revenues of US $ 303 billion, which is only marginally less than the gross domestic product (GDP) of Denmark (Blas 2013). The size of these companies makes it difficult to effectively influence these firms. Even if financial penalties would be much higher than they are, these types of firms could easily overcome the financial damage. However, as these firms continue to grow the risks of economic or market failure the moment their profitability dries up, are increasing. These firms might literally have become 'too big to deter' (Blas 2013; CEPS-ECMI Taskforce 2013; Martijn 2013).

Extralegal Enforcement as a Solution? Naming and Shaming and the Threat of Reputational Damage

In situations in which formal legal enforcement is lacking impact, 'naming and shaming', either by public authorities or by nongovernmental organization (NGOs) or the media, is often suggested as an alternative. Adverse publicity is often expected to reinforce public enforcement by inducing extra reputational sanctions on top of formal sanctions. The consequences of reputational damage in terms of financial harm due to loss of clients and stakeholders, and in terms of loss of personal status and prestige on the part of members of the business elite through shame and stigma, often exceed the damage resulting from formal sanctions. Therefore, the threat of adverse publicity may also deter firms from offending, and thus prevent corporate offenses. The wish to maintain a good reputation with clients, local communities, NGOs, and the media is a major motive for compliance and even for beyond-compliance investments by firms that depend on their reputation.

Public supervising agencies increasingly attempt to make use of the power of publicity by issuing press releases when a firm is prosecuted or by publishing sanctions naming the offender. Also, NGOs attempt to generate public condemnation of damaging corporate behaviour. The oil industry has been a frequent target. Greenpeace, for example, has repeatedly campaigned against oil drilling in the Arctic and oil spills in Nigeria. A recent, very powerful media campaign targets Lego for collaborating with Shell by using the popular 'everything is awesome' tune from the Lego movie as a background for Lego figures slowly drowning in oil.[11] It caused Lego to terminate the marketing contract with Shell, although Lego stated that it was unfair to involve the company in a conflict between Greenpeace and Shell. But

[11] https://www.youtube.com/watch?v=Ci4I-VK9jew.

it was no coincidence that Greenpeace targeted Lego, and not Shell. NGOs have to carefully select those targets that will create maximum public impact. Campaigns against firms with a close relation to end consumers and a 'feel good' factor, such as Lego, attract more media attention than campaigns against the firms that are unknown to the general public. This implies that the shipping industry is not the most likely target to attract media attention. Although some firms in the shipping industry are among the world's largest firms, they are often largely unknown to the general public and not as engaging as firms with direct end consumer contact. This goes for oil-producing companies, but even more so for the trading and shipping industry. The Dutch-based firms Vitol and Trafigura are the world's largest independent oil trading companies. Even though their operations and profits largely exceed those of more familiar Dutch companies such as Heineken and Philips, most people have never heard of them (Martijn 2013; Vink 2011). Only 9 of the 20 largest traders are listed (Blas 2013).These commodity traders have been characterised as 'the biggest companies you have never heard of' (Reuters 2011). This is one of the reasons why the threat of negative publicity and naming and shaming have less potential as a compliance-enhancing mechanism in the international shipping industry than in other, more consumer-oriented business sectors.

Not only does the character of the businesses in the shipping industry contribute to a low level of public condemnation of the dirty oil business, another relevant factor is that media coverage of environmental crime is often limited (Kraft et al. 2011; Lynch et al. 2000), particularly when it concerns environmental damage that is difficult to visualise, such as long-term damage to the ecosystem. Media coverage also depends on the availability of images, a condition that is often hard to realise in the case of oil spills at sea. Read (2011) reports how the Deepwater Horizon oil Spill allowed 24 h of live coverage of oil spewing out of a pipe into the Gulf of Mexico. When BP capped the well, the images stopped, and media coverage, which had risen to 40 % of news time, quickly declined (Read 2011).

The limited media coverage (and resulting lack of public condemnation) of the blending and dumping of dirty oil substances is also a result of active publicity management by the shipping companies. Corporate public relations (PR) strategies are often directed towards neutralising the harm of and denying responsibility for what they frame as 'accidents'. Media accounts of chemical spills as 'accidents' rather than as 'chemical crimes', which not only makes these spills appear less serious but also leaves the question of responsibility untouched (Lynch et al. 2000). In the Probo Koala case, for example, Trafigura Chairman Eric de Turckheim was interviewed on BBC Newsnight. With a charming French accent, De Turckheim stated that the Probo Koala carried 'a mixture of water and soda' blended in 'a routine operation'. He also stated that the incident was an accident for which not Trafigura, but the small Ivorian waste processor Tommy Company was to blame.[12] The civil damages that Trafigura paid to the inhabitants of Ivory Coast were not to be seen as compensation or an act recognising responsibility, but as an act of corporate charity. Another illustration of corporate PR strategies attempting to minimise reputational damage can be found in the way oil corporation Exxon prepares for an oil spill in its 'crisis preparation plan', as documented by Coll (2013) in his corporate biography of Exxon, 'Private Empire'.

[12] http://www.youtube.com/watch?v=tQBS82kFQjE&app=desktop (Accessed 30 June 2014).

Ten pages of this plan are devoted to the physical recovery of the spill, five pages to environmental protection, and forty pages to media strategies. The plan contains 13 standard press releases, in which the blanks can be filled in with the information about the event. For example, Exxon's standard press release in case of 'Public fatality':

> We are greatly saddened by this tragic event and express our deepest sympathy to the families of those affected. We are working with [APPROPRIATE AUTHORITIES] at the site (.)(Coll 2013)

In the case of criminal charges, Exxon's standard reaction is:

> We believe there are no grounds for such charges. This was clearly an accident.(Coll 2013).

In addition to attempts to avoid blame in the media, firms have been known to actively resist negative publicity by employing legal means. In the Probo Koala case, Trafigura employed the legal firm Carter-Ruck, specialised in reputation management. In 2009, the British newspaper, The Guardian, obtained a confidential document—later known as the Minton report—that contained information on the harmful nature of the waste as well as the relationship between the waste and the health problems of people living near the dumpsites in Ivory Coast. Subsequently, Carter-Ruck issued an injunction against The Guardian that prevented the paper from publishing the report. Moreover, The Guardian was issued a so-called super-injunction banning all mention of the injunction. About a month later, however, a member of parliament scheduled a parliamentary question revealing the existence of the injunction. The Guardian informed Trafigura's lawyers that it intended to report about this parliamentary question. Immediately thereafter, Carter-Ruck warned the newspaper that if they would report about the parliamentary proceedings, they would breach the injunction and would be held liable. The Guardian, therefore, decided to publish a front-page story saying that—a first in British history—a newspaper was prevented from reporting about parliamentary proceedings. This story was picked up by the Internet community and within a matter of hours details about the injunction and the Minton report were spread around the Internet, urging Trafigura to lift the injunction (van Wingerde 2015).

Even though the resistance to negative publicity could be interpreted as an indicator of the value that firms attach to their reputation, there are also authors who question the importance of a good reputation in contemporary corporate circles, and claim that corporate offenders are less and less sensitive to the threat of reputational damage. In his book, 'The Death of Corporate Reputation', Jonathan Macey (2013) distinguishes between corporate reputations—the public image of companies—and personal reputations carried by individuals in business communities. Although scandals may harm corporate reputations, these may be less damaging for the reputations of the responsible individuals. A person, who is labelled a corporate criminal in the outside world, may be regarded as a moneymaker or a problem solver in the inner business circle. As for legal sanctions, although a criminal conviction may still carry a stigma, this is no longer the case for civil or administrative enforcement or settlements, as these have been imposed so often that people have become used to those and they no longer form a blow to one's reputation. An illustration of this is that the executives of Trafigura, Claude Dauphin, who spent five months in an Ivory Coast jail, and Eric de Turkheim, are still primarily associated with wealth and power in the business press, and have remained in leading positions within Trafigura for years after the Probo Koala affair (Hoffman 2014).

Final Remarks

One of the most significant concerns of our time is how to avoid irreversible damage to our natural environment and how to ensure sustainability for future generations. This chapter focused on oil-related pollution by seagoing vessels as one of the issues that is extremely harmful to the environment, yet for a longtime lacked political and public awareness. Not only do marine fuels contribute significantly to the levels of particulate matter in the atmosphere, a problem that increases when the fuel is blended with waste, but also the illegal dumping of waste oil at sea is an ongoing problem.

In the first part of this chapter, we described the scope and nature of the dumping of oil at sea and the mixing of waste with HFO. This description showed that while these activities may seriously affect flora and fauna at sea and threaten our air quality, they are extremely lucrative as well. The high costs of legal waste management and the relative ease and impunity with which waste materials can be mixed with HFO or dumped represents a significant criminal opportunity structure (van Daele et al. 2007; White and Heckenberg 2011). Consequently, oil is frequently blended with harmful waste and waste oil is dumped in the sea.

We then highlighted several complexities that make detection and enforcement of the pollution by seagoing vessels extremely difficult. Due to the nature of the products involved, inspection authorities face serious difficulties in detecting the dumping of oil in sea and the blending of waste with HFO. Moreover, sanctions are often limited to financial penalties, which lack impact because the firms involved in the oil and shipping industry are among the world's largest and financially powerful.

Although naming and shaming is often presented as an alternative when enforcement is lacking impact, we argued that in the oil and shipping industry it might lack effectiveness because of the limited media coverage of oil-related environmental incidents, the lack of 'end consumer' contact, and the fact that the companies involved may be able to effectively neutralise negative publicity.

We, therefore, conclude that the need to design innovative prevention strategies is increasingly apparent. Rather than relying on detection and enforcement at full sea or in countries in the Global South, where enforcement capacity is often lacking, a more fruitful approach is to prevent the export of hazardous substances from the Global North in the first place. First steps have been taken: for example, in the aftermath of the Probo Koala affair, the Dutch authorities took measures to ensure that ships unloaded polluted substances at the nearest port. This prohibits ships from keeping the waste on board to off-load it only where this is as cheap as possible and proper treatment is not guaranteed.[13] However, in order to be successful, such

[13] Letter to Parliament, Dutch Minister of Environmental Affairs, 15 January 2015, IENM/BSK-2014/264196.

a strategy requires cooperation and information exchange between various national and international enforcement authorities as well as clear rules with regard to the substances that qualify as waste or recyclables. This ideal is still far from the reality in most countries, in which the economic benefits of the global oil trade often result in a lack of political will and power to act against its damaging side effects.

References

Amnesty International and Greenpeace Netherlands. (2012). *The toxic truth: About a company called Trafigura, a ship called the Probo Koala, and the dumping of toxic waste in Côte d'Ivoire*. London: Amnesty International and Greenpeace Netherlands.

Andersson, C., Bergström, R., & Johansson, C. (2009). Population exposure and mortality due to regional background PM in Europe—Longterm simulations of source region and shipping contributions. *Atmospheric Environment, 43,* 3614–3620.

Blas, J. (14 April, 2013). Commodities: tougher times for trading titans. Financial Times. http://www.ft.com/cms/s/0/250af818-a1c1-11e2-8971-00144feabdc0.html#axzz3Pf3jAsEa. Accessed 23 Jan 2015.

Camphuysen, C. J. (2010). Declines in oil-rates of stranded birds in the North Sea highlight spatial patterns in reduction of chronic oil pollution. *Marine Pollution Bulletin, 60*(8), 1299–1306.

Camphuysen, C. J. (2012). *Olieslachtoffers op de Nederlandse kust 2011/2012*. Texel: Koninklijk Nederlands Instituut voor Zeeonderzoek.

CEPS-ECMI Task Force. (2013). *Price formation in commodity markets: Financialisation and beyond*. Brussels: Center for European Policy Studies.

Coll, S. (2013). *Private empire: ExxonMobile and American power*. New York: Penguin Books.

Crist, P. (2003). *Cost savings stemming from non-compliance with international environmental regulations in the maritime sector*. Paris: OECD/Maritime Transport Committee.

de Buck, A., Smit, M. E., Faber, J., & van Grinsven, A. (2011). *Blends in beeld. Een analyse van de bunkerolieketen*. Delft: CE Delft.

Gibbs, C., McGarrell, E. F., & Axelrod, M. (2010). Transnational white-collar crime and risk: Lessons from the global trade in electronic waste. *Criminology & Public Policy, 9*(3), 543–560.

Grontmij. (2010). *Trends in lozingen op de Noordzee. Evaluatie VluVerO dataset 1992–2008, onderzoek in opdracht van Rijkswaterstaat*. Rijswijk: Dienst Noordzee.

Helmers, R. (2002). *Statistical analysis of oil pollution data (1992–2000) from the Dutch part of the North Sea*. Amsterdam: Centre for Mathematics and Computer Science.

Hertz, N. (2003). *The silent takeover—Global capitalism and the death of democracy*. Amsterdam: Pandora Pockets.

Hoffman, A. (24 March, 2014). Trafigura Founder Dauphin Exits as CEO for Medical Treatment. http://www.bloomberg.com/news/articles/2014-03-24/trafigura-co-founder-dauphin-exits-as-ceo-for-medical-treatment. Accessed 30 May 2015.

Jernelöv, A. (2010). The threats from oil spills: Now, then, and in the future. *AMBIO, 39*(5–6), 353–366.

Kagan, R. A., Gunningham, N., & Thornton, D. (2011). Fear, duty and regulatory compliance: Lessons from three research projects. In C. Parker & V. Lehmann Nielsen (Eds.), *Explaining regulatory compliance: Business responses to regulation* (pp. 37–58). Cheltenham: Edward Elgar.

Kraft, M.E., Stephan, M., & Abel, T. (2011). *Coming clean, information disclosure and environmental performance*. Cambridge: MIT Press.

Lagring, R., Degraer, S., de Montpellier, G., Jacques, T., van Roy, W., & Schallier, R. (2012). Twenty years of Belgian North Sea aerial surveillance: a quantitative analysis of results confirms effectiveness of international oil pollution legislation. *Marine Pollution Bulletin, 64,* 644–652.

Lynch, M. J., Stretetsky, P., & Hammond, P. (2000). Media coverage of chemical crimes, Hillsborough County, Florida, 1987–1997. *British Journal of Criminology, 40,* 112–126

Macey, J.R. (2013). *The death of corporate reputation. How integrity is being destroyed on Wall Street*. New Yersey: Pearson Education Inc.
Martijn, M. (2013). Over deze Nederlandse oliereus is nog nooit een Kamervraag gesteld. *De* Correspondent, 3 December 2013. https://decorrespondent.nl/438/Over-deze-Nederlandse-oliereus-is-nog-nooit-een-Kamervraag-gesteld/13471128-bc330511. Accessed 27 Jan 2015.
National Commission on the BP Deepwater Horizon Oil Spill and Offshore Drilling. (2011). Deep water the Gulf oil disaster and the future of offshore drilling. Report to the president. http://www.gpo.gov/fdsys/pkg/GPO-OILCOMMISSION/pdf/GPO-OILCOMMISSION.pdf. Accessed 7 Feb 2015.
National Research Council. (2003). *Oil in the sea III: Inputs, fates and effects, committee on oil in the Sea, U.S. National Academy of Sciences*. Washington DC: The National Academies Press.
Read, C. (2011). *BP and the Macondo Spill, the complete story*. Houndsmills: Palgrave MacMillan.
Reuters. (2011). The biggest company you never heard of. Special report. http://www.reuters.com/article/2011/02/25/us-glencore-idUSTRE71O1DC20110225. Accessed 27 Jan 2015.
Shumsky, T., & Kent, S. (2014). Big commodity traders gain clout. Trafigura, Vitol among a handful of firms shaking up markets. The Wall Street Journal, http://www.wsj.com/articles/big-commodity-traders-gain-clout-1404933074. Accessed 27 Jan 2015.
Snider, L. (2010). Framing e-waste regulation: The obfuscating role of power. *Criminology & Public Policy, 9*(3), 569–577.
Spapens, T., Bruinsma, M., van Hout, L., & de Jong, J. (2013). *Vuile olie. Onrechtmatig verwerken en mengen van olieproducten als vormen van milieucriminaliteit. (Dirty oil. Blending of dirty oil products as a form of environmental crime)*. The Hague: Boom Lemma Publishers.
Thornton, D., Gunningham, N. A., & Kagan, R. A. (2005). General deterrence and corporate environmental behavior. *Law & Policy, 27*(2), 262–287.
van Daele, S., Vander Beken, T., & Dorn, N. (2007). Waste management and crime: Regulatory, business, and product vulnerabilities. *Environmental Policy and Law, 37*(1), 34–38.
van Erp, J., & Huisman, W. (2010). Smart regulation and enforcement of illegal disposal of electronic waste. *Criminology & Public Policy, 9*(3), 579–590.
van Wingerde, C. G. (2012). *De afschrikking voorbij: Een empirische studie naar afschrikking, generale preventie en regelnaleving in de Nederlandse afvalbranche. (Beyond deterrence: Deterrence, prevention, and compliance in the Dutch waste industry)*. Nijmegen: Wolf Legal Publishers.
van Wingerde, C. G. (2015). The limits of environmental regulation in a globalized economy. Lessons from the Probo Koala case. In J.G. van Erp, W. Huisman, & G. Vande Walle (Eds.), *The routledge handbook of white-collar and corporate crime in Europe* (pp. 260–275). Oxford: Routledge.
Vidal, J. (2009). Health risks of shipping pollution have been 'underestimated'. The Guardian. http://www.theguardian.com/environment/2009/apr/09/shipping-pollution. Accessed 6 Feb 2015.
Vink, J. (2011). *'Het gifschip. Verslag van een journalistiek schandaal.' (The Toxic Ship. Report of a Media Scandal)*. Amsterdam: Prometheus.
Vollaard, B. (2013). *Preventie van illegale olielozingen op de Noordzee. Een onderzoek naar strategisch gedrag van scheepvaartverkeer*. Tilburg: Universiteit van Tilburg.
White, R. (2008). Toxic cities: Globalizing the problem of waste. *Social Justice, 35*(3), 107–119.
White, R., & Heckenberg, D. (2011). Key vulnerabilities and limitations in the management of hazardous waste and its disposal: A checklist assessment tool. *Journal of Environmental Protection, 2*(9), 1257–1263.

Prof. Judith van Erp is a professor of public institutions at Utrecht University, The Netherlands.

Prof. Toine Spapens is a professor of criminology at Tilburg University in Tilburg, The Netherlands.

Dr. Karin van Wingerde is an assistant professor of criminology at Erasmus University Rotterdam, The Netherlands.

Index

T. Wyatt (ed.), *Hazardous Waste and Pollution,* DOI 10.1007/978-3-319-18081-6